T0200010

THE ENERGY SWITCH

How Companies and Customers Are Transforming the Electrical Grid and the Future of Power

PETER KELLY-DETWILER

Prometheus Books
Guilford, Connecticut

Prometheus Books

An imprint of The Rowman & Littlefield Publishing Group, Inc.
4501 Forbes Blvd., Ste. 200
Lanham, MD 20706
www.rowman.com

Distributed by NATIONAL BOOK NETWORK

British Library Cataloguing in Publication Information Available

Library of Congress Cataloging-in-Publication Data

Names: Kelly-Detwiler, Peter, 1960– author.
Title: The energy switch : how companies and customers are transforming the electrical grid and the future of power / Peter Kelly-Detwiler.
Description: Lanham, MD : Prometheus Books, [2021] | Includes bibliographical references and index. | Summary: "The way humans produce, distribute and consume power will be cleaner, cheaper, and infinitely more complex within the next decade. In The Energy Switch, leading energy industry expert Peter Kelly-Detwiler looks at all aspects of the transformation: how we got here, where we are going, and the implications for all of us in our daily lives."— Provided by publisher.
Identifiers: LCCN 2020049921 | ISBN 9781633886667 (cloth ; alk. paper) | ISBN 9781633886674 (epub)
Subjects: LCSH: Electric power consumption—United States. | Electric utilities—Environmental aspects—United States. | Energy industries—Environmental aspects—United States. | Energy development—Environmental aspects—United States. | Energy policy—Environmental aspects—United States.
Classification: LCC HD9685.U5 K394 2021 | DDC 333.793/20973—dc23
LC record available at https://lccn.loc.gov/2020049921

♾️™ The paper used in this publication meets the minimum requirements of American National Standard for Information Sciences—Permanence of Paper for Printed Library Materials, ANSI/NISO Z39.48-1992.

To my wife Julie, who knows enough about what I do to know she wouldn't want to do it, and whose love and laughter are the limitless and treasured source of energy that keeps me going.

And to my fellow professionals in the sustainable energy arena, working tirelessly and passionately to create a better legacy for future generations you will never know. Your efforts lift me up and continue to inspire me.

CONTENTS

1

THE BIGGEST TRANSITION
IN THE HISTORY OF THE WORLD

On a chilly April morning in 2019, I drove from my home about an hour southwest to Fall River, Massachusetts, an aging mill town located on the shores of Mount Hope Bay. A hundred or so people and I boarded a ferry with about 100 or so other people, and we navigated our way through the stiff whitecaps driving down the bay. As we got closer, all eyes were focused on two behemoth 500-foot-tall cooling towers soaring above the greenish hulk of what remained of the now-shuttered 1,600 megawatt (MW) Brayton Point coal-fired power generating plant.

During its glory years, the plant had generated 20% of Massachusetts' electricity supply, supported hundreds of well-paying jobs, and provided $4 million in annual tax revenues to the town of Somerset.[1] At the same time, though, it was also a huge source of air pollution[2] and carbon, burning through a 50,000-ton barge load of coal—imported from as far away as Colombia—every four days[3] and covering local neighborhoods in a thin film of particulate soot.[4]

I had been to the site the previous year to write a story about the new plans to repurpose the facility and met with officials from the Commercial Development Company (CDC), who had bought the 307-acre site, including the environmental liabilities, from the power company for just $8 million.[5] During that visit we toured the plant, walked the dock where the coal used to land by barge, and—the highlight of the visit—went inside one of the cooling towers.

On my initial visit, CDC Vice President Steve Collins had outlined to me the company's plans for the site, which involved site cleanup and environmental remediation, salvage of material for scrap, demolition of the existing infrastructure, and preparation for its repurposing. It was the repurposing part of the story that had drawn my interest. The site, with

The eye of Sauron—inside the cooling tower. *Photo courtesy of the author*

its deep channel and proximity to highways and the open ocean, was a near-perfect location to support the logistics associated with the emerging multibillion-dollar offshore wind industry. Brayton Point had another significant characteristic as well: a 1,600 MW interconnection to transmission lines and the region's power grid. In that sense, Brayton Point was an ideal electric outlet for the offshore wind industry to plug into when the first facilities came online.

Brayton Point's conversion from the old to the new power grid is a near-perfect symbol for the enormous multitrillion-dollar energy transition happening in power grids across the planet, driven by societal changes and technological advances. It's a transition that is occurring with startling speed.

EXPLOSIVE CHANGE

We were far from alone out there: dozens of pleasure craft were clustered as close as they could get to the red marker buoys delineating the approach limits. Lights blinking, two orange Coast Guard launches and a police boat herded us back away from the site. Across the bay, traffic on the I-95 bridge came to a halt several minutes before 8:00 a.m. Meanwhile, we huddled at the rail, coffee cups warming our hands, and watched expectantly. For most, it was an opportunity to watch the towers blow up. For me, it signified the transition from the old to the new and, yes, the prospect of the explosion was undeniably alluring.

Right on the hour, as if to make the transition more pronounced, a loud boom announced itself across the water, and black puffs of smoke emanated from circular black lines at the base of each tower where charges had been placed. Almost immediately, both towers began to settle and then buckle as they fell. The entire collapse took no more than five seconds, leaving behind an enormous dust plume that blew across the bay and obscured the bridge, a huge pile of rubble, and extraordinarily ill-timed and costly $570 million investment.

The implosion. *Photo courtesy of the author*

Those cooling towers had taken four years to build and were only seven years old, a victim of changing times. It wasn't renewables such as the rapidly expanding offshore wind industry that were responsible for Brayton Point's death knell; rather, it was cheap fracked natural gas that flooded power markets and elbowed dirty and more expensive coal-fired power generation out of the way. But just as gas (with half of coal's carbon emissions and limited particulate matter) pushed coal off the table, gas will inevitably be shouldered aside as well. Natural gas plants are far from carbon-free. And in the sunniest and windiest locations of the United States, renewable solar and wind facilities are already driving down market prices and forcing the closure of some gas-fired plants. In other areas, utilities are shutting down coal-fired plants, bypassing gas-fired plants, and moving straight to a combination of renewables and batteries.[6] The signals pointing to a carbon-free grid are becoming increasingly visible, even though many challenges remain.

A TRANSITION THAT DWARFS THE INDUSTRIAL REVOLUTION

As society pours more carbon into the atmosphere, the planet is heating up, and it is becoming increasingly obvious that the biggest global challenge of our day is to slow down and limit that change so we and our children can live within a habitable climate.

As a significant contributor of carbon emissions, our energy systems will have to rapidly evolve to meet the challenge. The energy transformation will ultimately involve tens of trillions of dollars of investment and impact every aspect of society. This transition to a clean energy future will dwarf the Industrial Revolution in scope and magnitude.

It's clear that a confluence of technological, ecological, economic, and political pressure is driving our power grids to become something entirely new and different. What's less clear is exactly how the grid will evolve and what differences will emerge in various regions of the country and, indeed, around the world.

MY QUEST

Having spent 30 years in the power industry and the last decade as an energy consultant, writer, and communicator, I wanted to see if I could shed some light on this revolution. I wanted to know more about what was likely to occur, to see if I could understand the breadth of this massive shift and tell its story. For years, one of my frustrations as an energy professional was that I had a day job on which I needed to focus intensely. I was a subject-matter expert, toiling in my narrow part of the field, but I never had the luxury of time to look around and understand what was happening in the various sectors of the electricity economy. I could not fully contextualize what I was doing within the bigger picture, and I could not easily find the resources that would efficiently help me to do so.

I wanted to scratch that itch to witness these massive changes, and tell the story of what it looks and feels like to have one foot planted in the past and the other in the future. My objective was to tell the story from a variety of perspectives, using site visits to various locations as the on-ramps to each chapter. So, in early 2020, I began contacting professionals across the industry to set up a series of visits in the coming months. I had trips lined up to visit a rancher in Texas who had put seven wind turbines on his land, go to California and meet the professionals installing electric vehicle (EV) charging stations across the country, fly to Europe and witness the process of joining the two halves of a 107-meter wind blade—the biggest in the world—and visit a mile-long battery factory in China.

My plans were dashed on March 6, when I arrived at the NEC battery facility in Westborough, Massachusetts. Upon entering, the receptionist asked me to sign an attestation that I had not been outside the country for one month. If I had, she said, they would not permit entry. Within a week, it was sadly clear that COVID-19 was sending the world into lockdown, a new and unfamiliar landscape.

My new challenge thus became how best to tell the story in the absence of those planned visits. A combination of phone and video calls largely helped solve that problem. At the end of the day, the narrative still exists, and my desire to tell the story has not diminished.

While COVID-19 has ravaged the economy and killed hundreds of thousands, it has not diminished the urgency in the race to decarbonize our energy economy. In fact, reports suggest it may have accelerated the effort. Some see COVID-19 as a clear indicator of what a calamity of global magnitude can do to society, and therefore an urgent warning regarding slow political action and its implications for the future risk of climate change.[7] If a virus that is not associated with a high statistical risk of mortality could bring global commerce to its knees, some warn, imagine what a permanently altered climate can do. One need only ask the inhabitants of Australia or California, who have been affected by calamitous fires in recent years.

THE RACE IS ON

The good news is that it has already become clear that a transition to a cleaner energy future is inevitable. It's already happening. Cheaper natural gas–fired generators and renewable resources such as wind and solar energy have already overtaken the electricity-generating landscape in many parts of the country and spelled the doom of numerous aging coal plants across the United States. The U.S. government agency charged with tracking such things indicates that since 2007, over 530 coal units—representing over one-fifth of the country's coal generating capacity—have been permanently shuttered.[8] In 2018 alone, 20 coal plants were retired. Even plants in the country's coal belt are no longer immune.

Few people would have guessed a decade ago that it would be possible for 40% of the total newly installed generating plants in the second quarter of 2020 to be solar plants. But they were. Even fewer observers would have believed that a single renewable energy developer would proclaim to investors that it would spend $1 billion on batteries in 2021, capable of discharging "enough electricity to power the entire State of Rhode Island for four hours." But that developer, NextEra Energy, did just that. Its CEO commented in early 2020, "By the middle of this decade without incentives, new near-firm wind and new near-firm solar will be cheaper than the operating costs of most existing coal, nuclear and less efficient oil and gas fire generation units."[9]

It is now clear that we cannot stop the transition to a new energy landscape even if we wanted to. The real issue is not whether the change will happen. Rather, the critical question is how quickly society can mobilize to bring that about and minimize the amount of future economic loss and human suffering. A corollary to that question then becomes how quickly we can increase society's understanding of both the need for this transition as well as the enormous economic potential that it may also bring, as an entirely new energy economy needs new technology, skill sets, and capital.

A multitrillion-dollar transformation will both create new fortunes and incinerate capital bet in the wrong places. We are already seeing that play out with the ascendance of Tesla in becoming the most valuable car company in the world while at the same time coal companies go bankrupt and oil companies write off billions in assets.

INVOLVING EVERYMAN IN THE CONVERSATION

One fascinating aspect of this transition is that average citizens are beginning to become involved in the conversation. People who until recently almost never thought about electricity except as a service they received in exchange for paying a monthly bill are now beginning to evaluate other options. Several years ago, I was sitting at a local watering hole on a harbor in southeastern Massachusetts, looking out on the water as the sun was setting. My attention was distracted when I heard somebody from the next table say "solar energy." I turned to my left to determine who could possibly be discussing renewables at a bar. A middle-aged man in an old T-shirt sporting flecks of paint was boasting to his colleagues that he had put solar panels on his roof and was paying almost nothing for his electricity. When the local painter installs a solar system on his home, it's clear that we have migrated from ideology and environmental concerns to raw economics.

It's not just solar panels. Electric and autonomous vehicles look certain to overtake gasoline-powered cars within most of our lifetimes. That's not just a result of government mandates. They are simply better, and a growing number of customers want them. For example, in 2019, the Tesla Model 3 was the ninth most popular car in the United States, with almost 155,000 sold.[10] Those cars will not only take electricity from the grid; they will also deliver energy to the grid when it needs it the most. Wind farms and solar arrays are popping up all over the landscape from Arizona to Vermont and around the world, and batteries are now being deployed to capture that renewable energy and deliver it when we most need it. Our refrigerators, air conditioners, and hot water heaters are starting to interact with the grid and create value, too. Supercomputers are finding new ways to make the grid and its equipment more efficient. Computers are also struggling to keep our energy grids secure from hackers.

For the first time since the 1970s Arab oil embargo, many Americans are becoming involved in the energy conversation, and that's the discussion this book is all about. It's vital we have this conversation now, because change is happening all around us and we still have a chance to affect it, accelerate it, and nudge it in the direction that creates the most value sense for each of us. The way we produce and consume power will be cleaner, cheaper, and infinitely more complex in the decades to come, and our lives will change in ways we need to be aware of now so that we can plan for the future.

Incumbent utilities, energy companies, car companies, and other large industrial players will need to rapidly adapt—or perish, as the customer asserts a newly discovered power. So too will the local mechanic, gas station owner, and electrician. We are in the middle of the most profound change in history with regard to how we move about, do our work, heat and cool our homes, and communicate with others. And, yes, there are the painters and their solar panels out there, but many of us have barely noticed, and few fully understand the implications and the opportunities.

It is truly the most exhilarating and occasionally unsettling time for energy professionals, investors, utilities, and ordinary consumers that we have witnessed in the last hundred years, in large part because of the pace of the change. Our political landscape is just beginning to recognize these trends and deal with them. Examples include the many countries that have committed to meeting the goals of the Paris Accord, or the aggressive energy plan touted by recently elected U.S. president Joe Biden—an agenda-setting statement of principles that could put this dynamic on steroids. Even as the debate heats up, real change is already happening on the ground in many states across the country.

INCREASINGLY, A QUESTION OF CARBON AND TIMING

Sitting here in late 2020, the climate news is bleak. The National Oceanic and Atmospheric Administration (NOAA) just posted news that June 2020 tied as Earth's third hottest on record, with the six-month January to June period as the "second highest in the 141-year climate record."[11] And there's not much time left to cut emissions significantly enough to limit the temperature increase to 1.5 degrees C and maintain a habitable planet.[12]

As Mary Powell, the former CEO of Green Mountain Power, told me in a conversation in the summer of 2020, "Basically—as Bill Nye says—the planet is on fucking fire." Powell is just one of many leaders who have been pressing for a more urgent response to addressing the mounting crisis, marshaling the tools and resources at our disposal.

Powell commented, "The technology exists. The solutions exist. What we need now is bold courageous action focused on action and innovation. We need big money going there, we need innovation. We need it all, and there's room for everybody." In other words, a massive and near-immediate transformation is necessary, and the time for incremental approaches is over.

This book is about the elements of that action and innovation. It's about the people, the business models, the technologies, the trade-offs, and the trends driving that pace of change in one of the most complex and connected cyber-physical ecosystems on the planet.[13] It's about the scope of the transition that has already started, but will continue—of necessity—to accelerate.

SO MANY QUESTIONS, SO LITTLE TIME

How do we harness these new technologies to accelerate transformation, and what technologies will be involved? How do we push the transition to go the last mile to decarbonize our grid in a way that is cost-effective and equitable? How will we use the growing fleet of clean electricity resources to clean up other sectors of the economy, such as industry and transportation?

I believe we will arrive at our future more efficiently and equitably if we have a clearer sense of what is at stake, the tools at hand today and the ones we will likely have tomorrow, the trade-offs we may have to make, and the size of the opportunities that lie before us. This book is an effort to contribute to that informed energy conversation. It has arisen out of conversations with dedicated energy professionals who are working in various sectors all across the grid. It is an attempt to connect the dots and help explain the accelerating evolution of the power grid. And it is an effort to highlight the steps that are being taken by companies and customers that will forever change the face of the electric power grid—the largest and most complex machine ever constructed by mankind.

2

HOW ELECTRICITY ACTUALLY WORKS

To comprehend how the electric grid is transforming, why this is happening, and the implications of this metamorphosis, it's important to understand the basics: what electricity actually is, and how it's generated, transported, and consumed. This can be a little dry, but it helps set the context for the story that follows.

At a very high level, the key concepts for this chapter are the following:

1. What electricity is and how it is generated
   a. Current and voltage
   b. The difference between capacity (sometimes referred to as power) and energy
2. How electricity is moved across the grid, from generator to customer
3. How electricity is consumed
4. How the entire ecosystem is regulated

The big picture—how electricity gets to the consumer. *"United States Electricity Industry Primer,"* U.S. Department of Energy

GENERATION

Most electricity, with the exception of geothermal energy, tidal energy, or nuclear power, is in some way, shape, or form, a derivative of solar energy. Coal, oil, or natural gas–fired generation? Those come from sun's rays that struck the earth millions of years ago, nourishing plants and animals that were buried, subjected to pressure and heat, and converted into hydrocarbons. Hydropower? Evaporation from the heat of the sun literally lifts millions of tons of water into the atmosphere that eventually falls as precipitation on elevated topography. Wind energy? A result of uneven heating of the Earth's surface by sunlight (with some minimal help from the Coriolis effect, which determines the general direction).

The key difference between fossil fuels and today's renewables is the concentration of that energy. Hydrocarbons are—in their basic essence—simply extreme concentrations of solar energy that were sequestered in carbon over a long period millions of years ago.[1]

The methods used to convert the various starting forms of energy into electrical power boil down (sometimes literally) to only a few approaches. Nuclear power, coal-fired or oil-fired (and geothermal) generating plants boil water at high temperatures, resulting in the expansion of water molecules and creating steam. The resulting pressure—as much as 75 to 100 times atmospheric pressure—spins a rotor with blades on it (similar to the way a breeze turns a pinwheel).[2] Those blades—often dozens of them in different concentric sets along the turbine axis, or rotor—are designed to extract energy from the working fluid, whose energy declines as it passes through each row of blades—spinning the turbine rotor while maintaining desired pressures.[3]

Gas-fired turbines function in a similar fashion, but without steam. Instead, they combust a mixture of compressed, superheated air with injected gas at extremely high temperatures to turn a rotor that is also equipped with blades.

In both cases, the energy of the working fluid, in the form of heat and pressure, is converted into a rotational force, spinning the rotor in the turbine. Electrical power is generated based on the principal of magnetic induction. In the turbine, magnets are located on the spinning rotor, which itself is surrounded by a stator (the part of the motor that doesn't move) housing coils of copper wire. When the rotor spins inside the stator, it creates a changing magnetic field that causes electrons to move through the wires in an alternating current. Higher working fluid temperatures (such as those in highly efficient gas turbines) result in higher efficiencies.

The maximum power output of a generating unit is referred to as "capacity," and is defined in watts.[4] One thousand watts are a kilowatt (kW), and a thousand kilowatts are a megawatt (MW). One thousand megawatts are equal to a gigawatt, and one thousand gigawatts is called a terawatt. That's the capacity (or power) part of the equation.

Energy is simply the amount of power over a given period of time (power × time). So, for instance, a kilowatt generated or used over the period of one full hour is a kilowatt-hour (kWh). If you had a 100-watt incandescent lightbulb burning for 10 hours straight, for example, you'd consume one kWh. When referring to home energy use, we are all accustomed to talking in kilowatts and kilowatt-hours. However, when one moves to a discussion of the power system, the conversation quickly migrates to megawatts (MW) and megawatt-hours (MWh).

To provide some context around these terms, the average U.S. home consumes approximately 900 kWh per month.[5] It would likely have a maximum instantaneous power demand of perhaps five to eight kilowatts.[6] A decent-sized grocery store might consume perhaps 300 kW at any given time, a small factory might consume 1–2 MW, while a steel plant could have 50 MW or more of peak demand.

On the generation side, your typical gas,[7] coal,[8] or nuclear plant[9] might have a "nameplate capacity rating" ranging from 500 to 1,000 MW. Wind and solar farms have traditionally been smaller than that, ranging from 10 to 20 MW projects to many that are now well over 100 MW, and some exceeding 500 MW.[10]

The last issue to address is the concept of capacity factor. That is simply the average percentage of the time that a facility generates power over a specified period of time (usually a year). A capacity factor of 100% would imply that a power plant is operating at its full rated nameplate capacity for each one of the 8,760 hours in the year. No plant operates at 100% capacity, because fossil-fired and nuclear plants have to be brought down on occasion for maintenance.

It is instructive to look at the capacity factors of various types of generating plants in order to get a sense of how the entire generating ecosystem operates. Nuclear units run most frequently, and nearly all the time, with high capacity factors. That is both because of generally low operating costs, but more importantly, the process of fission is such that it is hard to dial back a nuclear plant. There is an effort to develop more flexible and modular nuclear units that will be both safer and more flexible (this will be addressed in chapter 15).

Coal plants were historically next in the mix, also running for many hours with high capacity factors. However, burdened by high carbon emissions and costs coupled with lower efficiency, coal plants are now being shouldered aside in favor of other resources such as gas-fired plants and wind and solar facilities.

Geothermal and hydropower plants generally don't factor into the conversation around the future grid evolution. Geothermal plants are small in number and make no material contribution, at .4% of overall generation.[11] By contrast, hydropower facilities make up a substantial portion of the generating fleet in places like the Pacific Northwest, but few new ones will be built because the best locations have already been exploited and the environmental costs are too high. The majority of hydro in this country (which still comprises approximately 6.6% of total generation and 38% of total utility-scale renewable electricity generation) can be dispatched when needed, but is dependent on precipitation levels.[12]

Natural gas plants are currently the critical swing player in the entire mix. They benefit from high efficiency and low-cost supplies of natural gas, with an increasing share of that gas coming from hydro-fracking that has driven gas costs down considerably in the past decade. Just as important are certain plant designs that are very flexible in the way they can be operated; however, they are normally less efficient. Gas plants can be ramped up or down to match constantly changing electricity demand, or to accommodate growing amounts of intermittent wind and solar resources, or both. This role will be discussed in greater detail in chapter 11.

There are also hydroelectric plants, many of which can store water behind dams to generate energy when needed. There are also pumped storage hydroelectric plants that pump water into reservoirs during periods of low-cost energy and then release water through turbines when energy is more valuable. These days, these facilities are increasingly being used to complement wind turbines and solar arrays.

These resources have energy stored in the fuel that feeds the facility: gas in the pipeline, coal in the pile, or water behind the dam, and are thus fully dispatchable—available at the grid operator's beck and call.

Then there are the growing twin renewable resources of wind and solar. Since each of these resources is highly dependent on the availability of local fuel (wind and sunshine), their capacity factors are highly dependent on location, as well as the underlying technology. For example, the capacity factor of solar projects increases as the panels become approximately 0.5% more efficient year over year, but some panels convert sunlight into electricity at 18% while others can achieve efficiencies of better than 22%. All things being equal, though, installing a panel in Arizona is going to yield almost twice the electricity as the same panel in Massachusetts. Developers are coming up with new ways to make solar plants more dispatchable so that they can ramp up and down according to the requirements of the grid.

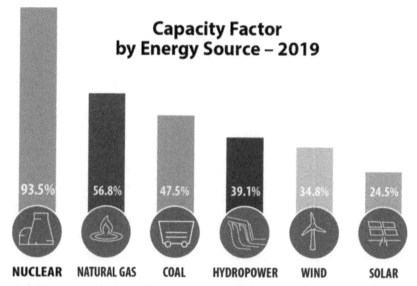

Capacity Factor by Energy Source – 2019

NUCLEAR	NATURAL GAS	COAL	HYDROPOWER	WIND	SOLAR
93.5%	56.8%	47.5%	39.1%	34.8%	24.5%

Capacity factors of various U.S. generating resources. *U.S. Department of Energy,* https://www.energy.gov/ne/articles/what-generation-capacity

Just as there are better locations for putting up solar panels, there are some areas in the country with more favorable wind resources—generally the states running from Texas north into the Dakotas, and off the shores of both coasts. Turbine height and blade length will greatly affect capacity factors (longer blades yield more energy, thereby increasing overall energy output relative to rated capacity). Overall average capacity factors are increasing rapidly as a result of improved technology.[13]

The challenge of moving to a carbon-free grid with the technologies we have today is essentially all about how we can integrate as much wind and solar as possible into the power grid. That brings a new resource into play—batteries (discussed later).

GENERATION DISPATCH CURVES AND ENERGY PRICING

A somewhat important topic for discussion on the generation side of the equation is the dispatch curve. This topic will be discussed in more detail later on, but it is worth briefly touching on here. Wholesale market electricity pricing is a function of three distinct elements: how much (capacity), for how long (energy), and where it's delivered (transmission constraints have an impact on localized prices).

The general principle of cost-efficient supply is that the market operator or utility calls on, that is, "dispatches," the least expensive generating resources first and then moves up the supply curve on the basis of cost. It's the classic "Economics 101" supply and demand curve.

In competitive wholesale markets, which cover roughly half the United States (the remainder still have vertically integrated utilities that own everything from generation to the wires and poles that serve your home), power plants bid in at a price that will yield a profit. They must at least cover their operating costs, including fuel if they are not renewables. If a specific facility is selected to generate energy, the marginal price it is awarded *gets paid to every other resource that has been selected to supply power at that time, irrespective of their operating costs.* Therefore, at any given time, the marginal resource necessary to meet demand is by definition the most expensive resource that has been dispatched to date. In competitive markets, the marginal resource sets the price for that particular moment.

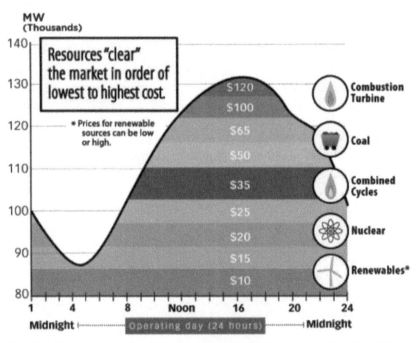

A typical dispatch curve (supply meets demand). *PJM Interconnection,* https://learn.pjm.com/three-priorities/keeping-the-lights-on/how-pjm-schedules-generation-to-meet-demand.aspx

The final, critically important element to understand here is that the renewable resources will always get dispatched first. Having no fuel to pay for, the marginal costs of these units are zero. Hence, renewable assets get called on ahead of every other asset in the dispatch stack. The other generating assets (especially flexible gas-fired generation) must then play the role of balancing between customer demand on the one side, and the growing fleet of intermittent renewable resources on the other.

In some markets, at certain times, the amount of renewable energy coming into the grid is such that energy prices can be very low. At times they can even migrate into negative territory if other conventional fossil-fuel resources cannot be backed down fast enough. In those instances, when prices often go negative, customers can actually get paid to use electricity.

In addition to the actual energy itself, there are other "products" sold in electricity markets. Capacity or resource adequacy payments, used in some markets, pay generating resources to ensure there is sufficient generating capacity to meet peak demand.[14] Grid operators and planners work to ensure there is sufficient generating capability to meet fluctuating demand, but sometimes there isn't. In mid-August 2020, for example, a scorching heat wave in California drove thermometers to new heights. That in turn created record electricity demand, largely as a consequence of increased air-conditioning load. Some generating assets failed to deliver expected output, with the result that the California grid operator actually had to implement a series of rolling blackouts for two days in a row.[15]

There are also markets for ancillaries such as spinning reserve (the ability to provide energy within a specific lead time, usually 10 minutes), frequency regulation (the ability to inject or absorb power from the grid near-instantaneously in order to maintain a desired frequency of 60 hertz), and other deliverables as well. There's even a market for a service known as "black start," which is the ability of an asset to "restore electricity to the grid without using an outside electricity supply," in the case of a widespread outage.[16] How these are defined and traded depends very much on the regional markets.[17]

Power plants supply these products to wholesale power markets, but customers can participate as well. By changing their consumption patterns in response to price signals or directives for the utility—behavior known as demand response—customers can often get compensated in much the same way generators do. In some markets, customer demand response assets can contribute 5% or more to the overall capacity mix, and that number is expected to rise greatly in the coming years.

TRANSMISSION

Once electricity is generated, it has to be transmitted to a user to be useful, and it can be moved in one of two ways. It can move in a direct flow from point A to point B, known as direct current (or DC), or it can oscillate back and forth in waves, or cycles, as alternating current (AC). The number of cycles per second is defined as hertz, and the North American grid operates at 60 hertz, or 60 oscillations per second.

There's also something called voltage, the force or "push" that causes the electrons to move within a wire or other conductor of electricity. It's the difference in electric potential between point A and point B. A commonly used analogy to explain voltage is a water tank with a hose attached at the bottom. In this case, the charge would be the amount of water in the tank (like energy in a battery), voltage would be the water pressure at the end of the hose, and electrical current would be the actual water flow. If you add more water to the tank, the charge would increase and the pressure on the hose—voltage—would also increase. Lower water pressure would result in less water flowing through the hose. Likewise, lower voltage results in less current (in essence, a smaller stream of electrons) flowing through the system. Higher voltage creates a "wider hose" and thus allows for the flow of more energy, which is why the electrical transmission system is characterized by high voltage lines.

Power plants generate at different voltages, typically between 2,300 and 22,000 volts.[18] However, since they all use the same transmission infrastructure, those voltage levels all have to be increased—or "stepped up"—to the same level prevailing at that location on the grid. This task is performed by transformers at substations typically located just outside the power plants. The majority of the roughly 180,000 miles of high-voltage lines within U.S. transmission networks are 345,000-volt (or 345 kilovolts, expressed as 345 kV) lines, although some transmit at levels as high as 765 kV.[19] These lines are typically made of aluminum or aluminum alloys.[20] Across the entire system, 5%–6% of the energy is lost, with about one-third of that in transmission and two-thirds in distribution.[21] Those losses come in the form of heat, and they increase as ambient temperatures increase.[22]

At the end of the line, though, many of our appliances function at 120 or 240 volts. Most outlets in your home are 120 volts with the exception of your electric stove, clothes dryer, and EV charger if you happen to have an electric vehicle. These are 240 volts. In order to deliver a flow of power that won't blow up your devices and set your home on fire, a series of transformers must again be employed, first to "step down" the voltage of the transmission lines at substations. These transformers typically reduce the voltage to enable subtransmission lines to move the electricity between 26 and 69 kV. These lines in turn connect to distribution substations that transport the electricity to where it is needed across lower voltage distribution lines, with voltages varying depending on whether the electricity is used in industrial or residential applications. The step-down transformers that serve residences are the round transformers one typically sees on telephone poles or on the ground in some neighborhoods. They enable the electricity to be delivered to the home at 120 and 240 volts.

With the exception of specially designed high-voltage direct current lines, DC power doesn't flow efficiently over long distances. As George Westinghouse and Nikola Tesla demonstrated back in the late 1800s, to Thomas Edison's chagrin (his was a DC system), it's far more efficient to move electricity over distances through AC. Thus, to date, the majority of transmission lines in North America are AC lines.

Large thermal power plants (steam and gas turbines) as well as wind turbines naturally generate AC, so they can be connected directly to the grid.[23] Solar panels and batteries, by contrast, generate DC and thus require an inverter to convert direct current to alternating current that can be integrated into the power grid.

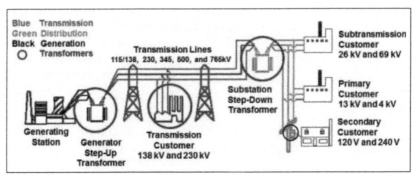

How electricity is transmitted. *"United States Electricity Industry Primer," U.S. Department of Energy*

The flow of electrons travels across that entire transmission ecosystem in a fraction of a second. Electricity is the only commodity on the planet that is both produced and consumed nearly instantaneously, and the demand is constantly changing at a local level, so it's critical that the entire grid is always in balance. One characteristic of electricity is that it seeks the easiest path from point A to point B. So, for example, if a power line is down, the electricity will move along the next easiest pathway as long as there is another line available.

TRANSMISSION FOR INTEGRATING RENEWABLES

Transmission becomes increasingly important in grids striving to integrate a growing level of renewable resources. In grids that are not well interconnected with neighboring areas, there is only so much renewable energy that can be accommodated before prices go into negative territory, or renewables are curtailed (with the energy wasted), or both. One can add batteries—and that is starting to occur with increasing frequency across the grid. However, a more efficient way to create a more robust power grid and integrate more renewables into the mix is to build transmission lines. With a more connected electrical network, the diversification of generators and demand at different times in different geographical locations means less capacity and storage is required.

In today's world, though, transmission is limited and at times it is constrained such that it creates inefficiencies. Not infrequently, there is not enough capacity on the lines to move the cheapest power from point A to point B where there may be high demand. Occasionally congestion results in the dispatch of higher-priced generating plants to serve load in a specific area. This may have the result of raising overall prices on the system.

In a perfect world, it might make sense to construct huge wind and solar farms in places like the middle of the country and the American Southwest and export that zero-carbon energy to the rest of the country. The reality is that opposition from those whose land and views would be affected—and may not directly benefit from this infrastructure—is often strong enough to make this approach politically impossible.[24] Consequently many of these projects, which can take over 10 years to permit and develop, often end up in the planners' graveyard. Nonetheless, there is so much potential value that numerous planned projects are still inching forward.[25]

There may be creative ways to solve this issue in some regions, for example by putting lines underground along existing rights of way such as highways and rail lines. The SOO Green Link is an effort to make this approach a reality. This proposed 350-mile, high-voltage direct current (525 kV) line will be laid largely along the route of the Canadian Pacific Railroad line. This $2.5 billion project will connect two different power grids, bringing as much as 2,100 MW of power from Iowa to Illinois.[26] Much of that energy is likely to be generated by wind turbines.

Advocates and planners are eyeing rights of way along rail lines and highways as potentially the easiest way to site new transmission lines, but a lot of political and permitting work will be required to turn this concept into reality.[27]

SIGNIFICANT INEFFICIENCIES
ACROSS THE ENTIRE CHAIN

Unfortunately, the physics of the power grid are such that energy is lost along the entire process from generation to transmission to distribution to the final end user. That equation is actually quite ugly. Expressed in British thermal units (BTUs), in 2019 an estimated 37.66 quadrillion BTUs of raw fuel were consumed to generate electricity. Approximately 22.9 quadrillion BTUs—60% of the entire energy input into the system—were lost in the conversion process, largely in the form of waste heat (when you see a cooling tower at a power plant, that's a large vertical expression of energy being wasted). Of the 14.77 quadrillion BTUs of gross electricity generated, 0.72 quadrillion BTUs were utilized within the power plants themselves, leaving 14.05 quadrillion BTUs. But then an additional 0.89 quadrillion BTUs were lost, largely in the transmission and distribution process.

At the end of the line, after all those inefficiencies, precious little of the original energy is actually left. Just over 13 quadrillion BTUs worth of electricity gets to the end user. The implication—from a climate perspective—is that most of our associated carbon emissions are a result of losses across the entire system.

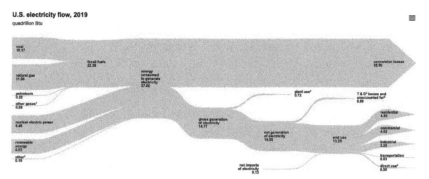

Losses across the system. *U.S. Energy Information Administration,* https://www.eia.gov/todayinenergy/detail.php?id=44436

Fortunately, there are ways to address this challenge. The first is to decarbonize the grid by utilizing as many clean energy resources as possible for utility-scale generation. The second is to bring more generation (such as rooftop solar and on-site batteries for storing the power) to the local level, so that transmission and distribution losses don't factor as heavily into the equation. And the third is to employ end-use technologies—lights, refrigerators, air conditioners, and computers that are more efficient. Replacing a 60-watt incandescent light with a 10-watt LED, for example, cuts consumption by 83%. Those savings ripple all the way back along the grid, reducing the impact of those losses that occur along the entire system. That is one reason why energy-efficient technologies are so critical in meeting the climate challenge.

THE REGULATORY CONSTRUCT

The final piece necessary to understand how the grid functions at a high level is the regulatory and oversight environment that ultimately creates the rules of the sandbox. There are essentially two critical regulatory layers here. The first is federal oversight, in the form of the Federal Energy Regulatory Commission (FERC). The FERC oversees the bulk power grid and interstate commerce, so it is responsible for seeing that markets are "just and reasonable."

The FERC defines wholesale market rules and continues to refine those over time, weighing in on a variety of issues from transmission siting to how batteries are compensated in wholesale markets. Texas intentionally does not move significant quantities of power across interstate lines, expressly so that Texas electricity can avoid being subject to oversight from the federal government.

The regional grid operators that oversee competitive wholesale markets in about half the country (except Texas) must create rules and oversee markets in compliance with the FERC's dictates. In all of these wholesale markets, generation is operated by independent entities that buy and sell in the marketplace.

The Northwest, Southwest, and Southeast regions are the principal areas of the United States that do not have competitive wholesale markets. In those areas, vertically integrated utilities hold sway, owning the generating assets as well as the wires and poles. Some areas are served by federal power agencies, such as the Tennessee Valley Authority or the Bonneville Power Administration, that also own the entire ecosystem of assets.

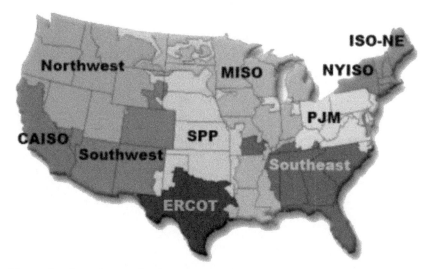

The regional markets. *Federal Energy Regulatory Commission,* https://www.ferc.gov/industries-data/market-assessments/overview/electric-power-markets

At the state level, utility regulators oversee policy and set the rules for the local investor-owned utilities (IOUs). In addition to the IOUs, there are also numerous electric cooperatives—generally rural entities that are owned by their members—and municipal utilities, owned by cities and towns, that report to local governments.

The best term to describe this bewildering patchwork of oversight, ownership, and responsibilities might be "Balkanized." That's the backdrop against which all of this change is occurring, and it makes for a very uneven level of regional progress and a highly complex energy conversation.

3

THE MOST VOLATILE
COMMODITY IN THE WORLD

A uthor's note: Events move quickly in this industry. As this book was in the final editing stages in mid–February 2021, a polar vortex invaded Texas. Temperatures plummeted well below normal, as two back-to-back storms pelted the state with snow all the way to the Gulf Coast. This unusual weather resulted in a calamitous situation with the Texas power grid, where gas supplies fell, equipment froze up, and nearly half the generation facilities went offline. Lengthy blackouts occurred as a result.

For the limited electricity supply that remained, power prices soared, pegging out at \$9,000/MWh for days. The facilities still able to produce power stood to make a killing, with some generators making a year's worth of revenues within a day or two.[1] Initial cost figures for the entire week were estimated at \$51 billion.[2] To put that in perspective, the *total amount* of electricity across the entire United States in 2020 (including delivery over the wires and poles) was just under \$391 billion.[3]

Approximately 29,000 residential customers, such as Kevin McAlpin (profiled in this chapter), were exposed to these high market prices, with many of them experiencing bills in the thousands of dollars. Even before the dust settled (and the snow melted), the first lawsuits began to fly and the finger-pointing began. On February 26, Griddy, the company that manages the electricity supply for his home, was forced by the grid operator to relinquish all of its customers for nonpayment.[4] The Texas polar vortex experience clearly illustrates just how rocky and volatile electricity markets can be.

Houston, Texas, resident Kevin McAlpin was away from his home on August 12, 2019. That was not unusual, as he travels a lot for business. What was more unusual on this particular August day was that McAlpin missed a notification on his smartphone from Griddy. That missed signal would end up costing him about $150 that day due to a spike in wholesale electricity prices caused by an extensive heat wave.

McAlpin's experience is what many customers could be looking at in the future. McAlpin can control how and when he uses electricity, cutting his overall bill and avoiding higher prices, depending on how vigilant he is.

The energy markets call this type of customer a "prosumer," which is a person actively engaged in managing their energy. In some states, prosumers can not only buy electricity from the market, but also sell it back to the market through various programs.

THE EVOLUTION FROM UTILITY MONOPOLY TO CUSTOMER CHOICE

McAlpin's ability to control his energy usage and price didn't just happen overnight. The evolution from utility monopolies providing electricity to every customer based on a tariff to an open market where consumers of all types can choose their supplier began in the late 1990s. Those early days of retail energy market restructuring in which pioneering customers could choose a supplier have grown to encompass many utility markets.

Today in competitive retail power markets across the United States, electricity consumers can choose to buy their energy from somebody other than the incumbent utility. Consumers in most of New England, New York, the mid-Atlantic states, Texas, and limited parts of California (but not the Midwest) can buy their electricity from a competitive supplier known as an energy marketer or retail energy provider.

Many consumers who buy from a retail energy supplier choose a fixed rate per kilowatt-hour for a specific period, for example a year. In order to manage their risk, cover their overhead, and make a profit, those retailers charge a markup for every kilowatt-hour they sell. If the retailer simply passes along market prices to customers like McAlpin, or many larger customers who are comfortable taking on that approach, they don't take on that risk, so they can charge a smaller premium.

In fact, in these same retail choice utility markets, *all* customers buy from a retail supplier, although they may not know it. Those who do not buy directly from a specific retail supplier have their demand aggregated together with all of the other consumers who choose not to participate. That aggregated consumption is then auctioned out to be bid on by—you guessed it—retail suppliers. It may be called basic service, default service, standard offer, or provider of last resort, depending on the state or region. But it's all supplied by a competitive entity.

FROM CHOICE TO PROSUMER

For most of us, electricity is an invisible commodity that we are unfamiliar with. Few of us have the savvy, interest, or time compared to the Kevin McAlpins of the world. We pay our power bill, and in some months we are surprised, while in others we are generally fine with it. This lack of engagement is partly due to electricity being an invisible and somewhat complicated product to understand, but it's also largely because the units on the utility bill reflect its costs of service, which are largely inscrutable to the average customer.

Imagine if we applied the electricity-buying experience to purchasing gasoline at the pump: What would that feel like? If I were setting up that scenario, I'd have you drive to the gas station twice over a given period, perhaps arriving once in the early morning, say 7:00 a.m., and once in the heat of day, maybe around 4:00 p.m.

I'd grab a roll of duct tape and would make sure to get to the filling station well before you arrived. First, I'd climb a ladder and cover over all the illuminated signs indicating the price per gallon so you wouldn't be able to get any initial indication of price. Then I would race to each of the pumps and ensure that I covered over any indicator revealing either the price or the gallons pumped. Better yet, I'd have somebody else pump the gas, so you wouldn't even know roughly how long you had been pumping the gas to estimate the volume.

Then I'd do nothing for five weeks, until finally I would send you the bills—far too late for you to change your behavior in response to any information contained on the invoice. The bill from 7:00 a.m. might be for $20—representing 10 gallons at a below-normal market price of $2.00. The bill for an identical quantity during the heat of the day could be for $900.00, representing the same 10 gallons, but time-differentiated to reflect the cost of supply later in the day. Same gas; wildly different price. Simply because of when you bought it.

That's kind of how many of us buy the billions of dollars' worth of electricity that flows through our homes, offices, and factories each day. We don't know how much we use. We don't know how much it costs us. We don't understand pricing, much less the volatility within pricing, so we just pay the bills and complain about them.

Fortunately, most of us don't get exposed to that price volatility. Somebody else—a utility or a retail supplier who sells us electricity—manages those risks, and most of us see a flat price. However, even though most of us don't think about price volatility risk, rest assured it is getting passed on to us as consumers.

A PROSUMER MODEL

McAlpin is not your average electricity customer. He has a lengthy background in electricity markets, and he knows all about pricing volatility and the fact that the price of electricity in Texas can actually go negative at times, or soar to $9.00 per kWh.

When Griddy founder and longtime acquaintance Greg Craig called him up a while ago and explained his new and innovative business model for proactively purchasing electricity for his home use, McAlpin enrolled immediately. For a $9.99 monthly membership fee, he can now buy electricity at the wholesale spot market price the same way the trading companies and big industrial customers do (Texas has a clearing price set every five minutes based on supply and demand, called the Market Clearing Price for Energy or MCPE). He fully embraces the technology to minimize his costs, and treats controlling his electric bill like it's a game.[5]

In a telephone conversation, McAlpin explained to me how the Griddy program works: "You download the app to your smart phone and then get pricing notifications. You get a bell alert when a price spike is occurring, and you can shut almost everything off remotely to avoid a high bill. When the spike is over, you return everything back to normal. If you want, you can do even more than that."

McAlpin gets his home's usage information from the utility smart meter on his house (you need to have a smart meter to make this work, but nearly all Texas customers have smart meters).[6] His smartphone app gives him the electric meter data, so he knows both his consumption and the market price for electricity at the same time. "The biggest users for me are heat and air conditioning," he commented. He precools his house in the morning when he won't be home, essentially building up a reservoir of cool air so he won't need to use as much air conditioning later in the day when there could be a price spike. "I will have it on really cold in the morning when the cost is cheap at a penny a kilowatt-hour. Then at noon I might just shut it totally off, which I can do from my phone using my app, if the price jumps a lot."

McAlpin manages his electricity usage patterns to coincide with the times of day when the power is either cheap or expensive in order to keep his monthly electric bill low: "I schedule dishwashing and clothes washing during low price times." In October and November, he noted, the price went negative for a short period during the day. In other words, he was paid to use electricity and pull power from the grid. McAlpin's response? "Hey, it's time to do laundry. You start to pay attention to things like your dryer. On the days you run the dryer, you see multiples of normal electricity costs because you ran the dryer that day."

However, even a seasoned pro misses a putt once in a while. The heat wave and resulting price spike on August 12 caught McAlpin off guard. More expensive and infrequently used generating units in the wholesale market got dispatched to meet the increased air-conditioning demand across the state. These plants bid in at high prices, resulting in punishingly high wholesale prices for a few hours—and McAlpin paid for it. Prices soared 36,000%[7] from their morning low in the pennies per kilowatt-hour up to $6.00 a kilowatt-hour for a brief period.[8] "I will be honest," he said, "The first day it happened, I was traveling and I didn't get the notification. That day alone probably cost me $150." But a seasoned pro also knows how to up his game, and McAlpin was determined not to get caught a second time. As the summer wore on, the price spikes continued to get worse, going as high as $9.00 a kilowatt-hour during certain hours. McAlpin commented, "I didn't feel the pain from these spikes, because I shut everything off." He ratcheted his electricity use down and continued watching the price signals. He estimates his total bill for the month of August was "probably $170 or $180. But most was that one day I missed the signal."

McAlpin recognizes that this approach is not for everybody. You must have the willingness and ability to change your behavior and use the data and tools available. However, he observed, "If you want to control your costs, then it's the way to go. I've probably saved $1,800 to $2,000 a year on my home electric bill."

While McAlpin's approach is way beyond that of the average consumer, his option to buy power from a nonutility supplier is not that unusual. Exposing customers to prices often results in behavioral changes, which in turn make the grid more economically efficient and reliable.

Griddy Chief Operating Officer Jason Huang noted that the Texas summer of 2019 showed that consumers did cut use in response to price. "We were able, in aggregate, from all of our Griddy members, to get a 25% reduction in usage during those peak times," Huang said to me. "We learned a lesson in terms of being able to provide more education and details, and how to communicate that to people. We saw some custom-ers reduce consumption 50%–60%, while others did nothing." In the long term, he views this as a potentially valuable tool in creating a more efficient power grid, because it creates a relief valve on the demand side of the equa-tion which previously did not exist. He also thinks customers will become more responsive as they become more familiar with what they can do to control demand. "We believe next time there's a peak, we should be able to reduce by 40% or 50% to remedy the supply-demand imbalance."

Griddy's approach is unusual, but it may spread over time.[9] The company's website indicates that for the four years prior to 2019, savings relative to the average rate hovered between 17% and 30%.[10] There's risk there, owing to the relatively inflexible nature of demand, but providing the customer with the real price of electricity is not much different from other markets—whether it's Uber's surge pricing, airline tickets, or gro-ceries—except in the order of magnitude. No other commodity can soar 500-fold or more in less than a day.

ELECTRICITY BASIC LESSON NUMBER 1: SUPPLY IS GENERALLY FIXED AND DEMAND ISN'T

While consumers aren't likely to become traders and most are not as so-phisticated as McAlpin at playing the electricity pricing game, prosumers are a growing phenomenon, and a basic understanding of how the power market works will become increasingly important.

There are some key features to grasp in understanding electricity markets. The first feature is that electricity is a commodity, just like grain, oil, or copper. However, that's where most of the similarities with other commodities generally stop. Unlike other commodities, electricity cannot be stored (although that's now starting to change).[11] Instead, electricity is a flow of electrons—moving rapidly across wires—and the only commodity on the planet that is both produced and consumed almost instantaneously. When that production-consumption, supply-demand relationship gets out of balance, serious problems, such as blackouts or power quality issues, can result.

The second feature is that electricity's value as a commodity depends largely on the quantity, timing, and location to which that flow of electrons is delivered. Those simple facts are critical in shaping the market environ-ment, and in creating—from a price perspective—the most volatile com-modity in the world.

The potential for huge trading wins or losses in those hot Texas August days was further anchored by two elements unique to the Texas market. First, the amount of excess power generating capacity above peak load—measured in available megawatts of capacity, or "reserve margin"—wasn't very large. In a system measured at 82,000 MW of available generating stations, the system only had 8.6% of reserve margin. So, the cushion meant to isolate the grid from a blackout or brownout in the event of failure of one or two of the larger critical generating plants was well below the 11% of the previous year, and significantly short of the targeted 13.75% margin that the Electric Reliability Council of Texas (ERCOT) planners typically call for.[12] This means that control room operators had far fewer available resources standing between them and the need to selectively cut electricity service to some areas or even risk a larger blackout.

Second, unlike every other competitive power market in the United States, Texas is not regulated by the FERC. Texas authorities have scrupulously taken pains for decades to ensure that no significant quantities of electricity move across boundaries to or from neighboring states, which would trigger regulation by the FERC. Texas is therefore free to set the rules of its regulatory market the way that it wants to. Not surprisingly, the state has opted to move in a direction quite different from its regional-market, FERC-regulated brethren. In other regulated wholesale markets, energy prices are capped at $1,000 per MWh by FERC rules.[13] Put in a way that will be familiar to the average person paying a monthly electricity bill, prices cannot exceed $1.00 per kWh for the actual energy portion of the bill.[14] Since everything is bigger in Texas, ERCOT market prices can potentially soar up to a specified limit of $9,000 per MWh (or $9.00 per kWh), roughly 300 times the average wholesale price over the previous half-decade.[15]

Let's compare the price of electricity to the spot market price for gold and gasoline to make the example of volatility clearer. The following figure depicts price ranges for each between 2015 and 2020.

Commodity	5 Year Low Price	5 Year High Price	Volatility Factor
Gold	$1,049/ounce	$1,514/ounce	<50%
Gasoline *	$1.83/gallon	$3.04/gallon	66%
Texas Electricity	($250.00)/MWh[i]	$9,000.00/MWh	Wow

*Wholesale commodity prices. MWh is megawatt-hour. *Electric Reliability Council of Texas (ERCOT)*

ELECTRICITY BASIC LESSON
NUMBER 2: WEATHER DRIVES DEMAND

Weather is what typically drives air-conditioning or heating usage, and that in turn drives power prices. More than any other type of usage, air conditioning has by far the biggest impact on its contribution to marginal demand. Every day, in every market, the common activities of humanity conspire to generate an electricity demand curve. It begins at midnight as most of us slumber. The machines in our factories are mostly silent, office buildings are dark, and most people and computers are in "sleep mode."

That cumulative demand curve stays relatively flat until the wee hours of the morning as the world beings to stir. Alarm clocks buzz, and lights, coffee pots, and hair dryers are switched on—each silently pulling a flow of power from the outlets and converting much of it into useful energy.[16] Commuters jump on electric trains, headed into their office buildings where they start up their computers and photocopiers, and the air conditioning cranks up.[17] Factories swing to life and electric motors begin their jobs of crushing, cutting, molding, shaping, heating, and sealing; that demand curve continues to climb, peaking at about 3:00 or 4:00 p.m. in most markets, and tailing off in the latter part of the day when people head home and the curve begins to head toward the quiet of the night.

As that relatively predictable demand curve evolves, the electricity supply must keep up with it, which means that some power plants are on all 24 hours and others must always be in "go mode," ready to turn on or off at a moment's notice to follow that curve. Every five minutes, each power plant bids into the wholesale markets at a price at which they are willing to sell, with each bid based on its own economic requirements. Those plants with higher operating costs will only bid in when they can cover their costs. They might only get used during a few hours of peak demand, and they need to cover fuel and operating costs.

Most days across the country, especially in the spring and fall, it's relatively easy to manage electricity supply and demand. Those seasons are the Goldilocks periods—neither too hot nor too cold—when demand is softer owing to reduced heating and cooling loads, and many power plants are taken down for routine maintenance. However, during weather extremes, and most especially the intense heat waves of summer, the entire grid comes under significant stress.

An extended heat wave generally begins to cause challenges to the power grid—and results in much higher prices on the third or fourth straight day of elevated temperatures.[18] The thermal mass—the glass, concrete, and bricks—of the city begins to heat up, and air conditioners across the urban landscape struggle harder and harder to maintain equilibrium. As they do so, they demand more electricity, which pushes the demand curve to new heights.

Coming into August 2019, the stage was set in Texas for a perfect pricing volatility storm. By early in the month, forecasters were warning of an extended 100+ degree heat wave for major cities such as Dallas and Houston. August turned out to be the second hottest on record, and on August 13 and 15 prices went stratospheric, bumping up against the $9,000 ceiling on both days for a combined total exceeding four hours.[19] That volatility was extreme: on the morning of August 13, energy traded for as low as $19 per MWh.[20] Later on that same day, that power price increased by nearly 475 times, up to the $9,000 ceiling. For those on the long side of the trades, whether the traders or those who owned power plants, it was a long-awaited windfall event.

Many of the larger industrial customers track the wholesale spot market prices the same way Kevin McAlpin does, with much the same pricing software used by traders that lets them view spot market prices in real time, so they can see the potential damage and alter usage accordingly. Those companies with backup generators normally deployed in times of grid outages fired up those machines at maximum levels to displace the need to buy electricity from the market. If the industrial sector had extra backup power capacity, they were able to sell it to the wholesale market and make a tidy profit.

MANAGING PRICE
VOLATILITY FROM THE INSIDER'S VIEWPOINT

The people in the world who most keenly understand commodity volatility risk are the ones who trade power on a daily basis. I wanted to feel what that energy conversation looked and felt like in the rooms where the money is made (or lost) and price risk is managed. So, I jumped on a plane to Baltimore on a surprisingly mild late January day to see some old friends—and make some new ones—at Constellation Energy.

Constellation Energy is a subsidiary of large energy holding company Exelon. Exelon owns a large fleet of power plants whose output it sells into various wholesale markets. Exelon also owns the retail and trading company, Constellation, as well as utilities Commonwealth Edison in Chicago, Potomac Electric Power Company in Washington, DC, Philadelphia

Electric Company, and Baltimore Gas and Electric. Exelon is structured so that its generation business unit competes in the wholesale markets, but its utility and retail business units aren't required to buy power from Exelon's generation business—they can buy or sell power from any number of generators or other trading companies, depending on what is best for each business.

I headed down to Constellation's relatively new and gleaming offices overlooking Baltimore Harbor on Dock Street, went through security, and was escorted up to the ninth floor. There I met my former boss, Mark Huston, president of Constellation Retail, the division of the company that sells power to residential, commercial, and industrial customers in competitive markets across the country. While the retail division must be very good at marketing and selling, it is critical to be exceptionally good at managing price risk.

As I was ushered into Huston's office, I was immediately struck by the view. Given Huston's title and responsibilities, I expected to be looking over the expansive view of Baltimore Harbor. Instead the office was focused inward into the central nervous system of the organization, with a view down onto the large carpeted trading floor populated by well over a hundred trading stations, each sporting a formidable array of computer screens. Each station was in turn grouped into a specific responsibility or geographic region.

It was an impressive view, somewhat like looking down on multiple booths at a trade show—with a lot more financially at stake. It was also surprisingly quiet and well ordered. Unlike the trading floors we see in the movies, with shouted orders and hands raised, there was none of that. Instead, people were at their trading station keyboards, eyeing data and typing in information, or standing in small knots having muted conversations, while data continuously flowed on the multiple screens around them.

MANAGING EXTREME VOLATILITIES STARTS WITH WEATHER FORECASTING

The trading floor's centrality reflects the critical nature of its role in the health and well-being of a large corporation that boasts the largest energy retail supplier in the country and also owns and operates 31,000 MW of generating capacity.[21] Of course, the trading floor buys and sells energy, principally gas and electricity. But what it really does is manage risk. A key component of managing price and volumetric risk is largely related to weather.[22]

At my request, Huston had arranged for me to meet PhD meteorologists Jonathan Mabry and David Ryan, as well as Patrick Smith, Constellation's manager of trading. Ryan explained that the role of the multiple meteorologists is to develop a view of the weather and how it affects short-term and longer-term trades for both electricity and gas, looking at everything from tomorrow and all the way through the coming summer or winter season.

Ryan commented, "I like to say that anything that deals with the weather will come through our desk, whether a heat wave or a cold wave that will freeze gas wellheads, or climate change." It all affects the markets, so the forecasting team arrives at their desks well before the rest of the office stirs, greeting the overnight team on the 24-hour desk as they head home in the morning.[23] At 6:00 a.m., the team formulates its view on the upcoming weather from that day to the next couple of weeks with "point forecasts" for major cities across the United States. Those local forecasts are critical because the urban hubs consume huge amounts of electricity and—being populated largely by office buildings—are extremely vulnerable to temperatures.

The weather group brings in data from numerous services, with the main sources being U.S., European, and Canadian models. Critical data include temperatures, humidity, and wind. Ryan explained, "Then we put together a market view. . . . It becomes about how we interpret biases, and experiences, and different forecasts in different situations." After that, Ryan said, they swing by the business units on the trading floor (each regional trading and business unit has meteorologists assigned to it) and offer detailed explanations underpinning their positions; "For example, we think it's going to be 95 degrees in New York and more humid versus another outlook, which could result in a big difference in price." Mabry added, "In the a.m. I will go to the ERCOT desk and provide our thoughts on our market view and how our view might be different from theirs. If I have a 102-degree forecast, I can give more color, saying there are guidance numbers as high as 105. Or, we might say, 'hey this cloud cover is coming in, so it could be lower.'"

The forecasts that Constellation's meteorologists rely on have become significantly more accurate over recent years as a result of increased computational firepower brought to bear by various government meteorological services. For example, the current National Weather Service computer that provides Constellation with one of its forecast sets has a calculating power of 8.4 petaflops—or 8.4 quadrillion calculations per second (roughly 10,000 times faster than the average computer we use at work and home).[24]

The question—once the initial weather forecast has been ironed out—is how it will affect demand and Constellation's overall position (long or short supply) for every location in which the company has exposure (purchases or sales). In wholesale markets, one can buy power for various timeframes and locations (in the Texas market, there are over 600 locations or nodes where one can buy power). This creates a very complex set of variables for managing positions and exposures and making the timely and informed decisions required to manage risk.

Not only do the traders rely on their weather team, they heavily utilize their analysts (affectionately known "quants"), who build and constantly update complex models that, among other things, aggregate consumption data by hour of every single electricity account being served in a particular location—for example, those 600+ Texas locations. Constellation parses the electric loads based on the load shape for residential, commercial, and industrial customers. These models will then help predict how sensitive to the forecasted weather that demand will be, and the company will forecast future demand (load) for each hour so that it can manage its position and make purchase or sale decisions down to the hour and specific location.

"With any type of volumetric risk, we also want to know what the extremes of those profiles are," Smith observed. "We might have a temperature forecast that's 101 in Dallas with a certain amount of humidity, but we also need to know for next week where the temps could really go. What does that mean for all the load we are going to serve?"

What that weather forecast boils down to is this, said Smith: "You think you are going to serve 100 MW of load in a market that can go to $9,000, so you might be worried that our forecasts could underperform and we could end up needing to buy supply at a very expensive time in the market. As trader, your job is mitigating risk and locking in profit margins using hedging strategies, so a bad forecast from my team makes the traders' jobs nearly impossible."

The other critical challenge is simple, but one that makes hedging and risk management particularly vexing. Customer demand follows a curve that may be shallow or steep, depending on the weather and the day of the week. By contrast, power purchases and sales don't work the same way; they are instead defined as fixed quantities over specific periods. In other words, they are rectangular and linear, compared with the rounded curves of customer demand.

"Most products we trade are on exchanges, typically power blocks that cover 7 by 8; 5 by 16, and then 2 by 16," explains Smith. What he means is, when traders are buying electricity, they purchase fixed volumes—generally a 50 MW minimum block for a given period. "7 by 8" means all seven days of the week, for the eight hours of off-peak, night-time demand from 10:00 p.m. through 6:00 a.m., while "2 by 16" are the two weekend periods of the same duration. In addition, traders can buy tighter ranges, such as a "super peak" product, during the later hours of the day when demand is greatest. Of course, this rectangular Lego block approach to buying power is not going to match the shifting customer demand curves dictated by weather. For every hour, the company's position is going to be either long or short, so part of what the position management team has to do is evaluate what they think the demand will be relative to the hourly electricity inventory they hold. The team can true up that position by purchasing power tomorrow (known as "day-ahead") to cover tomorrow's forecasted discrepancies. Then during the day of actual consumption, the trading team can buy or sell each hour in each location to balance supply versus load in what is known as the real-time market.

Smith summed it up this way: "In our business you are never perfectly hedged, and in every hour, you can be long and short in the same day . . . you're really trying to reduce your exposure. You have a fixed block, but your load is going to get higher in the higher-priced hours. You develop a view of 'do I want to be long? But what if wind comes in?' Then you don't want to be long." In other words, it's a high-risk game with a lot of unknown variables that can torpedo your best plans and sink your profitability in a heartbeat.

A TRADER'S VIEW OF TEXAS IN AUGUST

While McAlpin was managing his utility bill as a prosumer, Constellation was playing the same game from a different viewpoint. The Constellation team was watching the hot August weather closely but was not all that alarmed. Certainly, the reserve generation margins were low in Texas, so there wasn't a lot of extra generation to cope with excessive demand if the heat caused a pop in consumption. But from a meteorological perspective, Mabry said, "That August wasn't that hot, compared to normal, and that's what's concerning" in the long run. August 2019 had some extreme days, but it was not remotely like the excessive and prolonged Texas heat event of 2011, when temperatures soared well above 100 degrees for days on end and brought increased power prices with them. In fact, July 2019 had been relatively temperate. "So from our perspective," Mabry commented, "we were coming from a period where it hadn't been that hot. Then we started having forecasts for temperatures of 103 degrees in August. What was surprising was how this low threshold of heat (for Texas) had a dramatic effect on the market and prices."

Mabry looked to the future and speculated: what if this type of volatility becomes more commonplace? "That's a little more concerning. Temperatures in the low 100s are not that unusual for Texas. If a few degrees above normal becomes extreme, then that's going to be an issue." Mabry said he is asked every year if this is going to be another 2011, and he noted, "We didn't have the same kind of summer as 2011, nowhere close to the same amount of heat or duration, yet the prices for the summer of 2019 came out to be about the same as the summer of 2011. So, we can't just say 103 is not that hot anymore."

WEATHER COMBINED WITH RENEWABLES IS INCREASINGLY AFFECTING SUPPLY

During our conversation, Smith added an interesting observation that struck me as particularly trenchant. In Texas, he said, the number of wind turbines is now so significant that you have to pay attention to how the wind affects total generation output. If more wind blows, the supply of electricity increases, typically resulting in lower prices. Texas wind is typically an evening and nighttime phenomenon, so it can change the volatility of the market, and especially when one can expect to see lower or higher prices. In the old days—only five years or so ago—Texas peak prices usually occurred around 5:00 p.m. when the air-conditioning load was cranking. Now, though, cheap wind power is often being dumped into the system in the early evening, so the peak prices now occur earlier during the day.[25]

Ryan added that until recently, forecasting was very demand-focused and nearly all about temperatures—when a cold wave or heat wave was coming. Today, though, when a large portion of the fuel-feeding power generation is in the form of wind and sunlight, "we have to place an increasing emphasis on renewables that we have to forecast for." Wind is a particularly difficult challenge, he noted, because "wind is a different animal than temperature. Wind is heavily driven by terrain and very micro. So, you need micro-analysis to do the forecast for wind farms, particularly out west, because it's influenced by terrain. For example, California is a sea breeze. Texas is a low-level jet." And all that output changes hourly, daily, and seasonally.

IT'S GOING TO GET EVEN MORE COMPLICATED

It's tough enough today, but in the years to come it may become even more challenging for three reasons. First, the quantity of weather-influenced renewable generation is about to increase dramatically. Let's stick with the Texas market to explain why. The current ERCOT generating fleet stands at 82,000 MW of capacity.[26] In the generation mix, as of mid-2020 it has 23,860 MW of wind turbines currently installed and connected to the grid, with as much as another 12,000 MW of wind generation expected to come online by 2022.[27] Meanwhile solar generation, which stood at 2,821 MW at the end of 2019, is expected to soar to as high as 11,700 MW by 2022. The result will be that weather forecasting on the supply side will become increasingly critical.

Second, customer behavior will continue to change. Texas is one of the only regions in the country exhibiting positive electricity demand growth—due largely to the economic boom from oil and gas fracking and the increased demand for power needed by the wells themselves. How electricity is used in future years will affect future demand. For example, investments in energy efficiency, rooftop solar panels, and on-site batteries will cut the need for new, centralized large generating plants. At the same time, electric vehicles hold the potential to greatly increase overall demand. It's not quite clear where these and other unknown variables will take us, but it is clear that consumer behavior is changing.

Third, the climate is already having an effect on weather and resulting energy demand. A few years ago, ERCOT forecasters began increasing reliance on using the trailing 10 years of weather data—rather than relying on 30-year data—because the climate had become appreciably warmer. The 30-year data the forecasters formerly used were biasing ERCOT toward underestimating temperatures. In other words, the world has changed. Smith indicated that the Constellation trading floor is now focusing on 10-year data as well, and observed that "the 10- and 30-year used to be closer." Mabry chimed in: "The top five warmest winters have all occurred since 2000, with the top four warmest winters in the last 10 years, and we project the winter of 2019–2020 likely to fall within the warmest two or three."

What this means is that we can expect to see less certainty and more volatility in the energy markets. Weather is harder to predict, and sunlight and wind patterns make renewable power output harder to predict as well. Ever-changing consumer behavior adds even more uncertainty.

The critical element to understand in all this is that power trading is a zero-sum game. For every winner, there must be a loser that sometimes pays out exorbitant prices. For those on the short side who have not covered their positions, or are consuming electricity without a supply contract to address price risk, the cost of "being naked" can be extreme. A few short hours of high-priced power can be more than enough to torch the wallet. Some commercial or industrial customers who had been buying low-priced power on the spot market for years were cruelly reminded of that fact in August 2019, and got the financial equivalent of being whacked in the side of the head by a two-by-four. With a nail in it.

4

THE GRID'S CENTRAL NERVOUS SYSTEM

The Control Room

Electricity is strange stuff. Modern civilization cannot live without it. It powers our homes, our factories, our increasingly digital economies, indeed our lives. Without it, society would quickly disintegrate into chaos. And yet, we only vaguely understand what it is, how we generate it, move it, or consume it.

The Merriam-Webster dictionary defines electricity as a form of energy "expressed in terms of the movement and interaction of electrons."[1] It can conceptually be thought of as a flow of electrons that travels from the point where it is generated to the point where it's consumed, racing across a network of wires at nearly the speed of light, and constantly seeking avenues of least resistance. It's a commodity produced and consumed near-instantaneously, and the supply and demand for electricity must be in balance, lest bad things happen.

As a consequence, somebody must be in charge of the entire electric ecosystem, to watch over the grid and ensure that enough electricity gets to the right locations at the right time. In much of the country governed by competitive markets, grid operators (called either independent system operators—ISOs—or regional transmission operators—RTOs) manage the entire bulk power system.[2] Grid operators function akin to air traffic controllers, orchestrating the flows of power across the regional high-voltage transmission systems, ensuring that supply and demand are always in constant balance and the grid frequency of 60 hertz (in the United States) is constantly maintained.

Grid operators are also responsible for looking out 10 years or more and undertaking the necessary planning to ensure that an adequate mix of generation and transmission resources will actually be in place to meet demand, both regionally as well as more locally (to serve specific demand pockets, such as large cities).

And finally, grid operators must ensure that the financial markets—where wholesale electricity is bought and sold—work as designed and that no entities exert undue market influence that can distort or manipulate prices. That involves both oversight and the very complex process of accounting for every offering transacted on the market—and there are a lot of them. It's not just energy (in the form of megawatt-hours) that is being bought and sold. There are other products such as "capacity" and "ancillary services" bought and sold by various market participants.

A VISIT TO THE GRID OPERATOR

I wanted to better understand what grid operators do, so on a cloudy day at the end of January, before the lockdown that affected control rooms across the country, I set off to visit New England's ISO (ISO-NE). I pulled off the highway at an exit in western Massachusetts where the control room is located, and drove into an industrial park where a uniformed member of the local police force took my identification. I waited as the steel sliding fence rolled to one side, and eased my car into a parking space outside the building.

Once inside the concrete, steel, and glass building, I met Eric Johnson, director of external affairs, who introduced me to his team and led me to a conference room where, oddly enough, the interior windows were shrouded with blinds. Johnson commented that those windows looked onto the control room, and that the blinds would be opened at some point later in the conversation.

Mary Louise Nuara, senior external affairs representative, directed me through a 49-page PowerPoint deck prepared for my visit, outlining the roles and responsibilities of the New England grid operator. The task ISO-NE is charged with is complex: overseeing 350 "dispatchable" generators totaling over 31,000 MW of generating capacity serving the region's 14.8 million people over 9,000 miles of high-voltage transmission lines.[3] These lines snake across New England, and also connect the region to New Brunswick and Quebec to the north, and New York to the west.

AN INCREASINGLY COMPLEX EQUATION
WITH AN EXPLODING NUMBER OF VARIABLES

This task of balancing power flows across a changing landscape of assets has always been challenging, but it has gotten much more so over the past decade as the entire grid has begun to morph into something new, rapidly evolving and, until recently, also unimaginable. For much of the past century, the generating plants on the grid were typically quite large, and fueled mostly by a mix of coal, oil, natural gas, and uranium. This meant that—by and large—they *could be turned on and off* (dispatched) in response to a signal from a control room, because they had fuel available in the form of coal piles, oil tanks, and the like.[4]

All of that has begun to change in recent years, creating unprecedented challenges. The first of these being faced by ISO-NE and many of its counterparts across the country is the fact that numerous conventional fossil-fired assets are either about to retire, or have done so recently. The recently shuttered 1,600 MW Brayton Point coal-fired generating facility is just one case in point. For ISO-NE, 7,000 MW of generation (a little under 25% of its total fleet of assets) has recently departed the stage or will retire in the foreseeable future.[5]

Those retirements lead us to the second major issue. The generating assets lining up to replace retired plants are mostly clean wind and solar resources. In fact, nearly 21,000 MW of new and mostly renewable proposed generation sits in the queue, waiting to be connected to the grid.[6] That pending mix of clean power is driven largely by aggressive state decarbonization legislative mandates and aspirational goals (for example, the New England Governors and Eastern Canadian Premiers has put forth a goal to reduce greenhouse gas emissions by 75%–85% over the next three decades).

If past history is a guide, only about 20%–30% of the resources in the queue will ever see somebody eventually cut a ribbon and throw a switch.[7] However, those renewable resources that eventually do get permitted—dominated by offshore wind (which represents roughly 65% of the resources in the waiting in the wings)—will be intermittent.[8] That means they will generate electricity when nature's fuel shows up, in other words, when the sun shines or the wind blows. The remainder of the time, they will stay idle. Planning for that intermittency will require the development of other new resources such as battery energy storage. It will also result in the operation of existing resources in new and different ways, with more frequent ramping up and down (cycling) and starting and stopping in ways these power plants were never designed for.

With the exception of a few large offshore wind farms, these new resources are also a lot smaller, which means there will be many more to monitor. ISO-NE currently dispatches about 350 generating assets. However, as of February 2020, it had 953 new resources lined up in its interconnection queue, waiting to potentially connect to the grid at some future date.[9]

THE CUSTOMER AS A NEWLY ACTIVE PARTICIPANT

The evolution of the supply-side resource mix is only one piece of the new puzzle, though. The customer is also playing a more active and unprecedented role in power markets. Once a passive "ratepayer" who typically interacted with the power grid only when dutifully paying monthly bills, today's customer is becoming increasingly like Kevin McAlpin, engaging with the grid in new and unprecedented ways, including investing in sustainable energy-focused technologies.

In fact, about 10% of new resources (3,100 MW) identified by ISO-NE in its planning documents won't come from generation at all. Instead, they will be manifested in the form of customer-sited investments such as energy-efficient technologies (after all, saving electricity through installation of more efficient lighting and other equipment is the equivalent of building new supply resources).[10] Investing in efficient energy-consuming technologies is also generally cheaper, and involves no added infrastructural investments, or affiliated losses on the transmission and distribution lines.[11] In general, though, these efficiency investments involve passive technologies that simply reduce overall electricity consumption. But what's really shaking up the grid these days are the new on-site technologies—such as solar panels and batteries—that allow the consumer to generate and store electricity for the first time.

Flying into Boston on the day before my trip to ISO-NE, I peered out the window on the approach to Logan Airport to see if I could identify rooftop solar installations. Indeed, there were dozens visible, ranging from small arrays on residential dwellings to a few very sizable solar installations located atop large commercial warehouses. As the costs of solar installations continue to fall and state-level policies continue to push adoption, that number of small distributed power plants will only grow larger.

And soon, an increasing number of batteries will be either married to these solar panels or installed for backup power, a development with its own planning implications. In Vermont, for example, the utility Green Mountain Power recently set up a 10-year Resilient Home program for 500 of its customers under which they could install two Tesla Powerwall battery units, providing roughly 16–24 hours of emergency electricity supply in the event of a power outage. The cost to the customer was $30 per month, or a $3,000 upfront payment.

Meanwhile, all the way across the country in California, customers plagued by wildfires and lengthy preemptive outages from the local utilities in order to avoid additional blazes caused by equipment[12] were expected to add as many as 50,000 residential home battery units in 2020.[13] As the costs come down and corporate business models become more efficient, that trend will spread quickly to other markets, and all grid operators will eventually have to deal with both on-site solar and battery storage. It is now conceivable that in the foreseeable future, grid operators and utilities will see hundreds of thousands of customers with solar panels tied to battery systems, displacing the need to generate and transmit huge amounts of electricity formerly supplied by the central grid.

Prior to the recent invasion of customer-sited solar, batteries, LED lighting, and other efficiency investments, total demand could have been expected to continue to climb slowly. In the 1970s, '80s, and '90s, demand growth was a relatively predictable 3% year after year. The challenge then was simply to keep up with demand by building enough new infrastructure. Today, though, grid planners must plan for *decreasing* net electricity demand, declining on the order of 0.4% per year. That's a new dynamic they haven't had to address before.

However, perhaps the greatest challenge is that these new customer-located resources such as solar panels and batteries are *invisible* to grid operators. They are so small, they pass under the threshold and don't have to be registered with the grid operators, and therefore the ISOs can only model their behavior in aggregate. Planners must essentially forecast what consumption *would have been* and then subtract the demand that does not materialize on the grid because it's being supplied by on-site solar instead. In other words, the ISO-NE forecast is "net of" these other assets, which increase in number every year.

As a consequence of these accelerating dynamics, operators are seeing things in the control room they have never witnessed before. The most interesting recent example in New England was the impact of rooftop solar on net regional demand on April 21, 2018, a Saturday. Typically, the total generation needed to meet demand is much lower during the nighttime than during the day. However, on that April Saturday, a new phenomenon occurred that the control operators had been anticipating, but had not yet seen: the net generation needed to serve New England in the wee hours of the morning at 3:00 a.m. was greater than the supply required at 3:00 p.m. That was because, in the afternoon, roughly 2,900 MW of electricity was being generated by nearly 157,000 decentralized miniature solar power plants on people's rooftops all across New England.[14] This energy was being consumed on-site or exported back to the grid, cutting into the amount of energy that would have been required from the central generation fleet controlled and monitored by ISO-NE.

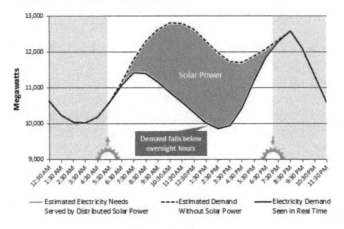

Historic Dip in Midday Demand Follows Record-High Solar Power Output on April 21, 2018
A sunny spring day pushed distributed solar output to an estimated record high of 2,309 MW at 1:30 p.m. and drove down electricity demand on the regional power system. In effect, New England consumers were using more grid electricity while they slept than in the middle of the day. (Data subject to adjustments.)

Even in New England, solar energy is affecting the grid. *ISO New England,* https://isonewswire.com/2018/05/03/a-regional-first-new-englanders-used-less-grid-electricity-midday-than-while-they-were-sleeping-on-april-21/

This task of integrating solar assets is already difficult enough in New England. In sunny California, the current epicenter of rooftop solar, it translates into a herculean task. In that state, over a million distributed solar arrays silently pour electrons into the grid every day. And it's going to get even more challenging, if the lessons from Hawaii are anything to go by. That tiny state has close to 82,000 on-site solar systems, a number that tripled in six years.[15] On the island of Oahu, the utility has 1,795 MW of centralized "dispatchable" generating capacity, as well as 524 MW of customer-sited solar that is invisible to the utility control room.

Like California, New Mexico, New York, Virginia, and a growing number of states, Hawaii has set for itself a goal of 100% carbon-free power within the next three decades, or earlier. In the case of Hawaii, planners believe that eventually fully 50% of the resources in that mix may be located on the customer side of the power meter. If and when that happens, managing the grid will be a task that is both centralized *and* highly decentralized, requiring new ways of operating control rooms and managing resources, and potentially requiring new grid architectures as well.

NEW AND HARD-TO-FORECAST SOURCES OF DEMAND

The customer-sited supply and storage resources are only part of the emerging picture, though. On the other side of that equation are the new technologies that may create vast new sources of electricity demand. These "beneficial electrification" technologies are being promoted by governments to address the specter of climate change and to push out dirtier fossil-fuel technologies. For example, efficient ground-source electric heat pumps utilize the thermal inertia of the earth to heat and cool buildings, displacing the need to burn gas or oil.

However, the real killer app may be the electric vehicle (EV). It has been estimated that a high adoption rate of EVs in the United States would cause overall demand to soar by as much as 38% by 2050.[16] Some markets, such as California—where policies support adoption of EVs, state-level incentives are generous, and charging infrastructure is also supported by strong policies and incentives—will see higher rates of adoption than other areas, but no area of the country will remain unaffected.

THE CARBON BATTLE: STATE POLICY VERSUS FEDERALLY OVERSEEN COMPETITIVE MARKETS

Ultimately, the resulting supply-demand dynamic will likely be highly regional, with each grid operator having to cope with its own increasingly idiosyncratic regional electric ecosystems. This will increasingly be driven by simple geography: where the most cost-effective renewable resources are located. After all, wind and sunlight are the available fuels for the renewable energy machines that harvest them. The same solar array in Arizona that yields a 29.1% capacity factor (that is, an average output of 29.1% around the clock) will only yield 16.5% in Massachusetts, owing to the latter state's more northern latitude and frequent cloud cover. Likewise, wind resources are remarkably abundant from the plains states down to the coast of Texas; not so in the southeastern United States.

Whether the resource is solar, or onshore or offshore wind, the same trends that are showing up in New England are making themselves known elsewhere across the country, with the grid operators more or less caught in the middle. Their mandates are to keep the lights on, and to "offer reliability at least cost," as one ISO-NE presentation put it.[17] While traditionally their role has not been to drive public policy, the grid operators find themselves increasingly engaged in this discussion as they coordinate and collaborate with each state and various industry stakeholders, and develop new approaches to address the mushrooming populations of new resources. That pressure on regional grid operators has grown more intense in recent years as mandates to operate competitive markets come into conflict with resources supported by both state- and federal-level subsidies.

That increasing tension came to a head in 2020, as the Federal Energy Regulatory Commission (FERC) has mandated that wholesale market operators force subsidized resources to bid differently into the market. State policy is now slamming squarely into federal regulation, with tensions rising as a consequence.[18]

The New York ISO (NYISO) believes it may have found a more efficient way to essentially sidestep this issue, simply by assigning a carbon adder to energy resources that bid into the market. More carbon-intensive resources would bid in at a price, including a specific carbon price adder, that would make them more costly than less carbon-intensive generators. If these carbon-intensive resources still get selected, the price paid for the carbon adder would then be reallocated to subsidize the further development of cleaner resources.

As a stand-alone entity, New York is the 11th largest global economy, emitting 0.5% of the world's carbon dioxide.[19] The Empire State's 2019 Climate Leadership and Community Protection Act (CLCPA) mandates that 70% of its electricity come from eligible zero-emission assets by 2030, with 100% being derived from these resources by 2040.

State policy affects the NYISO market, so, early on in the conceptual process, NYISO recognized that it would be helpful to engage the regulators, collaborating with the New York Public Service Commission to develop an appropriate framework. I had an opportunity to discuss this concept in the winter of 2020 with NYISO's CEO, Richard Dewey, and the way he framed the concept was simple.[20] The cheapest resources (weighted to include the societal portion of carbon) get picked first, so the overall market selection mechanism remains unchanged: "You still come up with the lowest cost dispatch. It's just that now the social cost of carbon is included as well." The ISO's software will determine both who pays and who gets to keep a specific portion of the carbon value. "If you emit a lot," Dewey said, you charge the adder, and then "you pay it all back."

How might that work in practice? Let's say a gas-fired resource could bid in at $40 per MWh, enough to cover its costs and yield a reasonable profit. If we assume a $3 carbon adder, it would have to bid in at $43. As a result of that higher bid, it might not "clear the market" and be selected, with cleaner assets meeting the gap between supply and demand instead.

If it were to clear the market, it would receive $43, and reimburse the $3 adder money that would then be funneled into a pot supporting cleaner generation resources such as wind and solar plants.

The intent of these market signals is to encourage more development of new clean energy supplies or improvements to existing resources to reduce emissions. Rather than forcing this through by mandate, or creating a patchwork quilt of regulations that pick specific winners and losers, Dewey said, let the buyers and sellers decide: "We think by using the raw efficiency of the market we can do that more cheaply."

Such an approach could be quite beneficial in both directing the market activities of today and helping steer the needed investment flows of tomorrow. It might also support developing technologies that are perhaps not viable today but may hold promise in helping society achieve a carbon-free grid. To get to a 100% carbon-free grid in two decades, Dewey commented, "you are looking at the necessity of developing technologies that are not commercially ready today. These might be renewable gas, hydrogen or large-scale carbon capture and sequestration, or something else." If entrepreneurs and clean energy investors know this additional reward is in the offing, it makes their investment thesis that much stronger. The next step in the process is for the NYISO to gain approval both at the state level and from its members. Then comes the potentially big hurdle: getting the concept past the FERC.

Dewey pointed out in our conversation that New York is not the only grid facing these problems, and that his state's approach could likely be applied elsewhere, because of its simple market-based logic.[21] "You have a single policy that can be evenly aligned with market design," he notes, one that addresses concerns of everybody and delivers the policy goal of a clean grid. In fact, Dewey's team has already been invited to present the concept at PJM (the Mid-Atlantic grid operator), and ISO-NE's CEO, Gordon van Welie, publicly characterized the plan as a "simple and easily implemented mechanism," asserting that the ISO has been in favor of a carbon price for quite some time. The problem to date has been that New England states have generally not been "in favor of pricing carbon emissions within the ISO-administered, FERC-regulated wholesale electricity markets," said van Welie.[22]

THE CONTROL ROOM

Regardless of how that conversation plays out, assets will still need to be monitored, coordinated, and dispatched from the control room, the central nervous system of the power grid. I had been so engrossed in my conversation with my hosts at ISO-NE that I failed to realize that something in the room had changed—the blinds behind me had been raised. I turned to the window as the lights in our conference room were dimmed to provide a better view of what was happening in the large, well-lit room below. There, a small number of semicircular work spaces populated the floor, each staffed by an individual, some talking on phones while others were sitting looking at their own screens or standing and scanning the massive 60 × 15 feet digital wallboard above them, which dominated the entire far wall.

Hundreds of rectangles were depicted in light purple. These were substations, where the power was directed from one point to another, "stepped" up or down to the necessary voltage in order to manage the flow of energy. Many of the rectangles were also accompanied by colored circles, each of which represented a generating plant, with its operating status and electricity output (in MW). These rectangular boxes were connected by lines showing the status of each element in the transmission system.

After a minute or two of perusal, and with some help from the ISO-NE team, the logic of the screen became clear. The map was an electrical representation of the entire New England region. So—as was to be expected—tie lines to New Brunswick were at the top right, while lines from Quebec were situated further to the left. To the lower right were the assets on Cape Cod. The upper left represented Vermont and Upstate New York, while the lower left depicted Connecticut and lower New York State. The digital map depicted the real-time power flows across New England as well as the active connections to neighboring markets. And situated high in the middle was a larger yellow number that barely moved, alternating from 59.99 to 60.00 and perhaps back down again: a precise and real-time measurement of the heartbeat of the system—its frequency in hertz.[23]

I was curious to know more about the human element, and the people responsible for managing the central nervous system of the New England region. Gleaming equipment and glowing screens are certainly impressive. But humans sit in the middle, and I wanted to know: what kind of person does that, and what do they actually do?

The ISO-NE team explained that first element necessary for running an effective control room is to hire the right talent. It is critical to bring on board the type of individual that thrives within structured procedures and protocols, knows when to take action, and consistently makes the appropriate decisions. Not surprisingly, many of the individuals hired for this role are ex-Navy nuclear personnel, professionals who are very comfortable with structure, protocols, and procedures. To ensure the right type of person is hired, many control rooms require applicants to take something called the PSP test, which evaluates aptitude, critical thinking, and analytical reasoning, as well as basic math background and word recognition.[24] ISOs receive recommendations from the testing company, as well as input from a psychologist who evaluates the results.

After extensive training, the selected hires join one of a small number of teams that work the control room, assuming a specific responsibility among the operators on the floor. That job may include contacting various generators, or scanning for the next potential problem and analyzing potential contingencies. Each team works a 12-hour shift, with handoffs occurring between each shift, for a number of days before getting a specified number of days off.

Previously, the default had been an eight-hour shift, but it eventually became clear that three shift turnovers per day increased the level of risk—as critical information about the current state of affairs must be handed off from one shift to the next in a tightly defined routine. Fewer handoffs results in fewer opportunities for miscommunication. The shift change requires the operators to develop a shift briefing for the incoming staffers to ensure that critical information is effectively communicated and the control room maintains "consistency of knowledge."

Those briefs may include a notification that a specific transmission line is out of service; that a certain line might have a temporary operating limitation, based on transmission operation guidelines; or it might discuss upcoming plans to take a line out of service. The entire turnover period may last 10–20 minutes.

Although it may run counter to intuition, since highest periods of demand occur during winter and summer when extreme weather can drive extremes in peak demand, the most time-consuming handovers generally occur during the spring and fall shoulder months. That's because it is during those months that generators take plants down for maintenance and transmission upgrades occur, with lines taken out of service. As a consequence, there may simply be more specific issues and anomalies to track during these shoulder periods.

There's also a constant focus on developing contingency plans to address various possible outcomes. For example, during my visit, a high-voltage 345 kV line linking New England to New York was out of service, with only one similar line remaining. What actions might the operators take in the event the second line were to fail? The operators already know what the backup plans entail and the actions they might have to take. They continuously rehearse and plan for various potential outcomes, even as they constantly watch the system map.

In a worst-case scenario, as recently occurred with the February 2021 polar vortex that hit much of the country, this may require resorting to extreme measures. In the case of Texas, freakishly cold weather resulted in an abnormal spike in electricity demand. At the same time, gas pipelines froze, cutting fuel supplies to power plants. Frozen instruments took other generators out of action, including additional gas plants, coal facilities, and a nuclear unit. Some wind turbine blades seized up as well, so at one point nearly 50% of the entire generation fleet was sidelined.[25]

As the crisis began to unfold, 2,600 MW of supply—enough to power 500,000 households—went unexpectedly offline in just 30 minutes.[26] The situation continued to deteriorate quickly from there, as more power plants became unavailable. The Texas grid operator, ERCOT, struggled to stave off a complete and catastrophic blackout that would have left the state in darkness for weeks or months. ERCOT's only recourse was to begin "load shedding" or cutting power to customers, eventually shutting off electricity to millions of Texans for days as they endured subfreezing cold temperatures and many struggled simply to stay alive. Some did not. Initial estimates suggested fatalities numbering in the dozens.[27]

Loss of that control oversight function for any extended period of time could result in adverse economic outcomes and equipment damage, so the successful performance of the grid operator control room is essential to the functioning of civil society. For that reason, ISO-NE maintains a near-exact replica of the main control room roughly 25 miles south in northern Connecticut, a backup that acts very much like simulators in airlines, and as close to 100% reality as one can get. The main control board looks just the same, and the entire facility must be capable of running at full functionality within 120 minutes. The Connecticut facility is also very helpful in running training exercises, creating a simulated environment so that operators can continue to train and develop responses to various contingencies that may occur on the grid.

THE HUMAN FACTOR

One individual who understands these issues and risks very well is Mike Legatt, CEO and founder of Resilient Grid, a company that develops information tools for grid operators and managers. A former principal human factors engineer at ERCOT, Legatt had to understand how control operators perceived and acted on information, and to build better information tools to convey that information efficiently and without overloading the human recipient.

In 2016 Legatt founded his own company to help grid operators and utilities deal with these issues in their control rooms. In late March, as the coronavirus reports were becoming increasingly dire, I gave Legatt a call to discuss the challenges utilities might be facing. This was just before the three dozen New York grid control room operators went into full lockdown mode at two redundant facilities outside Albany, New York.[28]

Legatt observed that a critical challenge the operators must deal with—even in the best of times—is the growing number of smaller grid-connected assets—many of them renewables or batteries. Each developer of these assets has its own version of securing and communicating data, which makes the situation more complex. The rapidly growing population of invisible on-site customer-located assets adds to the challenge because the operators don't know how many there are, where they are located, or how fast that population of connected devices is growing.

Artificial intelligence (AI) offers the potential to help, but Legatt commented that the capabilities we need don't exist today. In addition, he cautioned, an overreliance on AI creates the potential to result in "out-of-the-loop syndrome," in which operators become so far removed from the decision-making process that if something does go wrong, it may be difficult to reinsert them into the loop and quickly gain the required situational awareness.

Legatt summed up the challenge: "What it really comes down to is that these people are trying to balance increasing amounts of data, changes, and threats. Unfortunately, as the complexity of the grid increases, we are moving more and more to a place where these operators need to be creative and collaborative, because it's harder and harder to build a procedure for every possible situation."

COMPLICATED AND
COMPLEX ARE NOT THE SAME THING

Legatt noted that "we call the grid a complex socio-technical system, because the most complex creatures on the planet built the most complex infrastructure, and that's what the grid is."

This evolution of the grid has fundamentally changed the nature of the control room operator's job. A couple of decades ago, Legatt said, if you were a grid operator coming in at the beginning of your shift, you could grab a cup of coffee and stare at the wall before the shift change. You could probably understand everything that was happening in about five minutes because there were fewer dynamics to keep track of. Not any more, commented Legatt: "Now it's the case that much faster dynamics mean what you saw when starting your shift may not be in any way similar to where things are in the middle of your shift."

As today's grid evolves, there's a growing challenge because it is "increasing drastically both in terms of complexity and complicatedness." What's the difference? Complicated involves many independent parts that make up the whole. Legatt posited the example of a car. With complicated, "If the suspension is bad, as you drive it's going to be really uncomfortable . . . but if there was a problem with the radio somebody else could fix the radio problem while you fixed the suspension problem. They were both independent." Legatt commented that complexity is quite different: "Complexity means that you have multiple pieces that are interconnected in ways, some of which you can see and some of which you can't. You've seen crime-scene dramas where they have the map with push-pins and threads running between them. To me complexity is when you grab the push-pin, and pull, and you don't know what the pins are going to pull along with them. Generally speaking, as humans our cognition, collaboration, and organizational culture bring a lot of complexity. On a physical system, increasing the amounts of interconnected components also increases complexity."

What types of problems might that create? Imagine, Legatt suggested, thousands of homes with price-sensitive smart thermostats that all react nearly instantaneously to changes in the electricity market. "When you hit the peak, all thermostats go up, and demand drops and prices fall. And then they start again. You have more assets that are harder to control, predict, or model."

Legatt cautioned that these types of unexpected occurrences can lead to even the best-trained grid operators becoming overwhelmed by what's going on and lose high-level understanding of the system, that critical "situational awareness." As the grid becomes more complex, operators may apply the wrong mental model to a decision, thinking the system is operating one way when in fact something entirely different is occurring. It may not be immediately visible, or immediately make sense, to grid operators watching these variances play out. He cited two such incidents that occurred with fires in California several years ago to make his point.[29] In those cases, the fires affected the frequency on the transmission lines, causing resulting faults. Inverters connecting hundreds of distributed solar facilities responded to the temporary change in grid frequency by "tripping" and going offline, disconnecting the solar arrays so that they no longer fed power into the system. Spontaneously that solar energy just disappeared, and then 15 minutes later came back online. But all that control room operators saw were inexplicable increases in load (in one case 1,200 MW and in the other, 900 MW) that apparently popped up and then vanished again.

LIFE IN THE TIME OF COVID

Even before the NYISO had put its operators in lockdown, Legatt and his team were already focusing on the potential implications of the coronavirus pandemic that was affecting the country. "In COVID-19 world, the environment has changed drastically," he commented. The job cannot be done from home for cybersecurity reasons. "If I could do it from my home computer and the bad guy got in, it's game over for the grid. What that means, though, is that when utilities are constantly trying to think through worst-case scenarios, most envision leadership sitting around a table, sitting near the control room, talking to operators while looking at the control room. But since the primary source of reliability of the system is the system operator, that means that their showing up is actually putting the grid at risk."

So now, all of the other individuals in the system—the directors, executives, and those in media relations who need to communicate and field questions—all have to stay apart from the operators. The only way they can maintain any type of situational awareness is "to remote in from home and see read-only views of lots of screens that were built for multi-monitor control rooms, and videos that were never designed to function on a laptop. So instead of eight people around a table staring through glass onto the control room video wall, you now have eight people at home logged into at least five different systems . . . so you have impaired situational awareness."

Legatt noted that going into lockdown is stressful. "There are fatigue issues," he commented, and they will become more pronounced over time. "These operators are locked in, and they may not see their families for a month. Or more. There's not only fatigue, but they're probably not sleeping as well, either. People who are introverted are probably able to handle this better than extroverts. For some operators, you may have deferred trauma because you didn't have time to prepare, and it's possible that families are at risk."

PREPARING FOR THE FUTURE

So what can we do to prepare for the inevitable challenges to come?

"The answer," Legatt said, "is actually very simple. We have to make a choice as a society as to how much do we want to be resilient versus how much do we want to be efficient. From the perspective of resilience, it's really about the idea that you can recover from an unexpected situation or—even better—take time to anticipate and solve the problem before it happens. If you have humans operating at 100% capacity, then you have little slack to adjust to these sorts of things."

Legatt stressed that it all really comes down to the culture. In high-reliability organizations, "people would take time to understand what other people are doing next to them, because in an emergency you could take over and also have better collaboration." He cautioned that this approach is very different than becoming compliance-focused, which often results in managing to a least common denominator where "the regulator owns your mission." Legatt went on to say, "Operating the grid reliably is now a multidisciplinary problem. Power systems engineers are critical and will always be critical, but one of the things we are finding is that your culture is highly predictive of your human error rates." In other words, poor culture increases the risk of bad outcomes.

The organization must recognize that collaboration and information-sharing are critical as the level of complexity increases. Otherwise, each actor may miss signals that—taken together—would have clearly indicated a looming problem. If the culture is not collaborative, the risk is that "no one puts the puzzle together until it's in the rearview mirror."

"This is entirely a human problem," Legatt observed, "and the solution is not to add another procedure or another engineering restriction that is actually going to make the problem worse. Really, to me, the organizational culture is a series of habits and expectation—if I do this thing, this is what will happen."

5

GO FIGURE:
TERAWATTS MEET TERABYTES

Several years ago, I went on a pilgrimage to Berkeley, California, to pay a visit to Cori. Cori is a supercomputer named after Gerty Cori, the first American female scientist to win a Nobel Prize. She's run by Lawrence Berkeley National Laboratory's (LBL) National Energy Research Scientific Computing Center (NERSC), and she can perform 30 petaflops, or 30 quadrillion calculations per second.[1] That's helpful if you are interested in studying the large-scale structure of the universe or modeling a supernova. It also comes in quite handy if you are one of the many scientists around the world trying to create new compounds that will lead to breakthroughs in the sustainable energy field.

PHYSICALLY, ELECTRICITY
IS ALL ABOUT MATERIALS SCIENCE

Whether we generate, transmit, or consume electricity, we're fundamentally dealing with physics and chemistry, and we are depending on various technologies for every aspect of that entire network. Employing better materials can help us improve those various underlying technologies utilized across the grid, and that's where Cori comes into play.

Part of her job is to help researchers around the world develop better technologies by creating new compounds—materials formed by chemically combining two or more different elements. In other words, Cori's task is to dive into the periodic table and create a list of dozens (or perhaps hundreds) of new, hitherto nonexistent virtual molecules that can help address a problem, often at the nanomolecular level.

Cori can zip through the periodic table. *Photo courtesy of National Energy Research Scientific Computing Center*

At LBL, I met Kristin Persson and Horst Simon, who explained to me how they were using Cori, and they arranged for me to tour the facility. When I entered the building where Cori was housed, I was immediately struck by two impressions: the interior of the facility was spotless, and it was loud. Powerful fans pulled outside air through a lengthy array of Cray computers linked to one another by cables behind each machine. As we walked along the row of Crays, my guide pointed out a glass tile in the floor through which springs were visible; they protect the facility from seismic activity. My guide then ushered me into an adjacent empty chamber even larger than the room we had just departed. This was to be where the inevitable and more powerful successor to Cori would be located.[2]

Another thing struck me on my visit: whatever Cori was working on was completely opaque to the human observer. Invisible it may have been, but potentially invaluable in accelerating the quest for a better energy future.

The beginning of the process starts with an applicant—an individual or company—that defines the issue they are trying to address and the desired properties in the compounds they seek. Is it a potential photocatalyst material that can enable artificial photosynthesis? Smoother glass for the front of a solar panel so that soot and dust will simply slide off and increase overall efficiencies? A more powerful and stable battery?

Persson explained that the process is about "using the data Cori produces to filter through how materials, chemistry, and structures correlate with some of those properties for specific applications." The requested searches can be quite varied, but they must be worth Cori's time, as well as the assistance of a researcher who helps define the search parameters and creates the necessary code to guide Cori.

Once that code is written, Cori then begins her quest, with her multiple servers running hundreds of jobs at the same time. She may spend days combining various elements in the periodic table to come up with a list of potential chemical compounds. Compounds vary, but they have in common certain properties defined by the "quantum characteristics" of the atoms of which they are composed.[3] They have differing physical properties such as stiffness, elasticity, density, hardness, and conductivity of light and electricity, which can be investigated virtually. Cori's job is to find the compounds with the best qualities for the particular task at hand.

DRIVEN BY CLIMATE CHANGE, GLOBAL INTEREST HAS GROWN

In a phone conversation with the team in the spring of 2020, the researchers indicated that as the focus on climate change has grown, so too has awareness and interest from researchers all over the world, who have access to much of the publicly available data at no cost. "That has caught on in the last two years," Lead Researcher Persson commented. "A year and half ago the website almost went down because we had so much load." Persson noted that NERSC was not originally designed to be an always-on data portal for the global research community, but the demand has become so huge that they have had to accommodate it. "We have pressure to stay up 24/7," she said, and commented that everybody expects always-on availability. "We have scientists in Australia saying. 'Where have you been the last two days?' Materials science is becoming a community vehicle." Demand is so great, in fact, that her organization is considering outsourcing the results to a searchable cloud application that would provide the necessary uptime.

Deputy Laboratory Director for Research and Chief Research Officer Horst Simon indicated that the facility now has over 120,000 users across the world, with 3,000 to 5,000 logging into the site every day. "We are one of the most popular material databases in the world. We are published, free, and known to produce trusted data. There are publications associated with all datasets that are peer reviewed."

During our conversation, I asked Simon if they had made any progress on filling the big empty cavern I had seen on my visit. As it turns out, they had. Plans are well underway to construct Perlmutter,a machine that will be three to four times faster than Cori.[4] The actual specifications were still in negotiation, but are expected to fall within the range of 60–100 petaflops.[5]

Simon indicated the new supercomputer would be physically completed by the end of 2020, and fully configured the following year. It's not just about more speed and power, however; the new $210 million NERSC computer will be equipped with an architecture geared more toward data-centric computing. Among other attributes, the machine will help address the rapidly growing interest in machine learning.

The information gleaned from Cori and her planned successor, Perlmutter, can lead to valuable results. Persson cited work the team had done for Duracell a few years back, where computations were able to identify "200 out of 100,000" possible compounds that eventually led to the commercialization of a new higher-density battery Duracell calls Optimum.[6] Persson noted that while the inquiry eventually led to a successful outcome, "it still took 12 years to commercialize it. . . . Even when we come up with really cool materials, the road to commercialization is long." Duracell still had much work to do to winnow down the 200 potential candidates to identify the optimal one. The company then had to create compounds that had not previously existed, a process that involved its own challenges, since there is no recipe for an entirely new molecule.

Unsurprisingly, battery chemistries are continuing to elicit a great deal of interest among the global research community these days. In particular, the quest is on to find a solid-state battery chemistry that is more stable (less likely to catch fire or explode), has a higher energy density (so that it can extend the range of electric vehicles (EVs), for example), and can offer more cycles during its lifetime.

IDENTIFYING THE COMPOUND IS ONE THING; MAKING IT IS ANOTHER

Persson observed that finding the next great material for a battery or any application, for that matter, can be accelerated, but cautioned that "materials science is a long road . . . in the end there is a lot of engineering going into the last couple of steps." Once the promising compounds have been defined, they must then be created, or synthesized, in the real world to determine how they actually perform. She described the process of synthesizing the new compounds as easily taking two years and, thus, is "one of the slow things, so we are working to understand synthesis better. We are getting so good at predicting the novel materials with specific properties, but now we need to understand synthesis."

Persson was optimistic about progress made in this area, given a recent increased level of interest and the coupling of robots to do the actual synthesis with project data. Artificial intelligence is now beginning to be applied to interpret results from the synthesis process, "coupling with computer data and making decisions on the fly as well. An automated loop will be part of the lab in the future."

Simon added, "We are now bringing automation we brought to computing to synthesis—that's now the slow step. That's the holy grail of coming up with novel materials." That's an area where robotics and the concept of "self-driving labs" is showing great promise in accelerating synthesis. The other key advantage of utilizing robotics and automated labs is that the experiments are documented in ways that do not occur when humans are involved in the process.[7] The documentation of failures allows the self-driving lab to continue to learn and improve.

Persson explained, "We traditionally do not report our failures in publications, but if a robot fails, we keep the data so it helps us better understand trends." One issue that has impeded our ability to learn is that the current system of academic publication is geared toward highlighting successes, but this means that society is not placing enough value in learning from failure. However, she noted that with automation, the ability to harvest useful data to interpret failures will help researchers better understand where to go. Persson noted wryly, "You can't train a model on only successes. There is no Journal of Failed Experiments, and nobody would publish in it anyway."

INFORMATION TECHNOLOGY WILL ULTIMATELY REMAKE THE ENTIRE GRID

It's not just in the manufacture of "better stuff" that information technology (IT) has the potential to revolutionize the grid. Compared to other industries, the IT revolution has come relatively late to the power industry, but it is about to come to the fore and will do so in a number of ways that will render the future energy industry vastly different from the current one. This infusion of IT will affect all aspects of the industry concurrently. At a minimum, it will:

1. Advance our understanding of how physical assets on the grid work, in order to help them run more efficiently or for longer periods of time.

2. Change the way we manage and operate power grids, as increased understanding of real-time conditions and operating costs, based on geographic constraints, will allow us to optimize grid operations in ways that are currently unavailable to us.

3. Revolutionize the way fleets of power-generating assets are managed. Supervisory cloud-based networks will increasingly interact with asset-level machines at the grid edge to optimize performance with the context of numerous datasets.

4. Alter the way energy consuming devices behave, via a vast network of sensors in our buildings and factories linked to real-time feeds related to pricing and grid conditions that will enable end-use devices to respond automatically within the context of a price environment.

ECONOMICALLY, IT'S ALL ABOUT THE DATA

At a basic level, the nation's power grid is a physical entity, supplied by over 7,500 power plants[8] and connected by 200,000 miles of high-voltage transmission lines[9] and millions of miles of distribution lines that deliver electrons to our homes, factories, and commercial buildings.

At the same time, the grid is one gigantic data exercise, one of the largest in the world, that is about to get a whole lot bigger. Part of the reason is that electricity is the only commodity that is produced and delivered almost simultaneously. It's entirely different from loading a product at a port in one location and delivering it to a single fixed location. Instead, the flow of electrons must be delivered to *all* consuming locations, all the time, in order to meet demand.

That demand is constantly fluctuating. If you flip the switch on your domestic 1,800-watt hair dryer (which might temporarily double the amount of power your home is pulling from the system), the grid must immediately accommodate that decision. Likewise, if a steel company starts up a 100 MW electric arc furnace to melt steel, the power grid must accommodate that change in behavior as well. As demand constantly fluctuates, that balancing of supply and demand must occur across the entire system at every moment in time, both at the grid level as well as the local level. That feat alone requires a significant quantity of information processing.

The data intensity increases when you add competitive wholesale markets to the equation, and prices are associated with volumes over time for various services that generators offer to the grid. There's a price for every megawatt-hour, in forward markets where one commits to buy and sell energy to be delivered in the future in various time-delineated blocks. Do you want to buy all 100 MW for every hour in 2021 to support your manufacturing? There's a market for that. How about just the month of July 2021? There's a market for that, too. What about weekends and holidays? Or weekdays just from 7:00 a.m. to 11:00 p.m.?

What if you want 100 MW for a specified period tomorrow, with that energy delivered to a specific location (zone)? The ability to provide generation within 10 minutes, or to ensure the grid has a desired frequency of 60 Hz? What about the ability to start up a generator after a blackout so the grid can get up and running again (known in the industry as "black start")? You can buy those markets as well. Energy trading floors exist to do just that. There's a market for all these activities, and all that data must be recorded, stored, and delivered.

Not every market is highly location-specific, but the experience of Texas suggests we may see more granular and sophisticated market structures in the future. The Texas "nodal" market calculates the cost of transmission from the generating plant to a specific point of delivery (node), with the goal of better reflecting the combined costs of generation *and* transmission constraints.[10] This approach, in which market prices are established every five minutes, is intended to make the system more efficient.[11]

This means that supply and demand auctions are occurring all the time at 8,000 locations, but such a market-efficient solution can only occur within a system that can accommodate huge amounts of data.[12] The total cost of implementing the Texas program was over $500 million, with the lion's share allocated to external contractors and software.[13]

That's not where the data complexity ends. Where competitive power markets reign, the world gets even more complicated. In these markets, the utilities bill for delivery charges over their wires and poles, while competitive retail suppliers charge for the physical energy delivered to the end user. So now there are two sets of data involved. The utilities typically bill based on volume, or the number of kilowatt-hours that flow over their wires, but they also charge a cost based on the highest level of customer demand during the month, which can represent up to 30%–40% of a utility bill in some parts of the country.

With smart meters, it's possible to know at the account level who consumes how much for any timeframe we care to measure, in near real-time, so consumers can now be charged an hourly rate for electricity based on the cost of supplying electrons for that period of time. On a very hot afternoon during the third day of a heat wave, when air conditioning is straining to keep buildings cool and the grid is overloaded, the real-time cost of using that hair dryer or operating that electric arc furnace might be very different. That's the game Kevin McAlpin is playing in Texas when he buys electricity for his home at spot market prices. Sometimes, the price is negative and he gets paid to use electricity. At other times, spot prices can be 300 times the historical average.

Truly, the entire electricity market is one enormous data play, as we combine 8,760 hours of annual energy data with different hourly (or even 5- or 15-minute) prices. That would seem complex enough. But today we are entering a whole new world in which prices are based on where one is located on the power grid, transmission constraints, and the condition and replacement costs of the surrounding electricity infrastructure.

DATA INTENSITY IS
EMERGING ON THE DISTRIBUTION GRID

With the advent of big data, which lets us measure just about anything anywhere—and store the data at a negligible cost—we may soon enter the realm of micro-pricing environments. We may see a situation in which prices reflect not only the costs of supplying a customer at a specific time, but also at a precise location. In some cases, such costs could vary considerably. To illustrate this, let's take two theoretical households, each of which consumes electricity in an identical manner. The only difference is their respective location: let us assume that one (Customer A) lives next to a substation, where the power from the high-voltage transmission line is stepped down to the lower voltage so that it can be carried across smaller distribution lines to the end-user. The other (Customer B) lives at the end of a 10-mile-long distribution line. Typical average distribution line losses are around 4%, but they increase with the length of the line, and are magnified as lines heat up.

On a cool spring day, the residence at the end of the distribution line might require 2%–4% more electricity to be fed into the system, even though the end use consumption is the same. Now, let's take an extremely hot day where each customer's usage increases—as does everybody else's on the system—as a result of increased air-conditioning load. In this case, resistive losses (analogous to friction losses) occur. As the electric loads increase, the wires heat up, resulting in line losses of up to 20%.[14] This condition worsens the farther down the distribution line a user is located. In this case, Customer B would require considerably more electrons (20% or more) be fed into the system to support exactly the same level of home consumption relative to Customer A closer to the substation. As a consequence, an investment in an energy efficiency measure could potentially be worth 20% more to the system, and might even forestall the need to add new infrastructure to the grid in the future. For these two homes, such disparate situations have always existed, but now we can measure them.

Of course, not every locational difference would be that extreme, but it's not difficult to overlay geographic information system (GIS) data onto a utility infrastructure map to learn two critical elements: 1) the avoided cost—in other words, investments that are deferred or avoided completely—of not having to make specific improvements to a piece of infrastructure (whether it be a transformer, substation, or larger distribution line); and 2) under what conditions, such as increased demand and line losses from a heat-stressed grid, such investments would most likely be necessary. For example, if we could limit demand during peak periods to as few as 10 or 20 hours a year, we might not have to build out expensive new infrastructure to meet that demand.

Once these elements are combined, it is possible to create a micro-pricing environment that links locational avoided-cost considerations with the time dimension to create hourly prices reflecting costs of supply. These prices could be significantly different for multiple locations across the grid. This pricing equation could stimulate an economic response to four essential questions: where is the energy consumed, how much energy is consumed, for how long, and when? Each of these has an impact on value and price and, taken together, represent an enormous amount of data.

One initial proof point of this concept is in Consolidated Edison's (ConEd's) service territory in the Brooklyn/Queens area of New York City. Here ConEd was faced with an estimated $1.2 billion price tag to upgrade a substation serving parts of these boroughs, especially during hot days when air conditioning demand was expected to drive new levels of peak demand for a small number of hours during the year.

Instead of this expensive project, the utility received approval from state regulators to invest up to $200 million in measures to cut peak demand, including batteries, fuel cells, energy efficiency, demand response (paying users to cut consumption during peak periods), and solar energy.[15] Initial auction results demonstrated that the value of demand response in the program was worth nearly five times as much as for other areas in the city.

This approach is being replicated elsewhere within New York State in an ambitious undertaking to change the way the power grid operates, with the goal of decarbonizing the grid while lowering overall energy costs. This approach explicitly calls for the utilities to establish distributed service platforms (DSPs), which enable the deployment of these new distributed energy resources (known as DERs, and also sometimes referred to as non-wire alternatives) to minimize the need for new large infrastructure projects.

This approach is intended to complement the functions of the grid operators by optimizing energy use and supply at the local level. A specific goal is to "develop and implement vibrant markets for distribution system products and services." These include "managing market operations and processes, and administering markets," identifying market rules, and designing and conducting auctions for DER, "facilitating and processing market transactions," as well as "measuring and verifying participant performance."[16]

In other words, there is an enormous IT play brewing at the very edge of the grid, on the distribution system, in potentially hundreds or thousands of locations. A recent New York State report on the subject highlights essential technologies necessary to support required functionalities, such as "geospatial models of connectivity and system characteristics, sensing and control technologies needed to maintain a stable and reliable grid, optimization tools that consider demand response (DR) capabilities, and the generation output of existing and new DERs in the grid . . . to be supported by a secure and scalable communications network."[17]

Implementation will not be easy, and the process will take years; however, if early economic benefits can be realized and New York stays the course, it is not hard to imagine a highly decentralized energy marketplace in the near future. This could well include localized micro-pricing environments that vary by time, depending upon available supply resources and infrastructure constraints as well as constantly changing demand. In September 2020, the FERC issued Order 2222 specifically to encourage the growth of such assets.[18] Within the coming decade, it is probable that projected growth in DERs will require billions of annual transactions. In other words: data, data, and more data.

Data and IT platforms can be expensive, but the prices of chip sets and connectivity are now so low—and continuing to fall—that it is not hard to imagine a fully automated distributed electricity marketplace of the not-too-distant future. In that world, flexible assets such as stationary batteries, EV, air-conditioning units, and LED lighting systems could "know" the benefits and costs of participating in the marketplace and respond accordingly. LED lights could dim, for example, while batteries in basements and in cars would know when to charge and discharge based upon local information. At some point soon, such interactions are not only likely, but inevitable.

GENERATING FLEETS GO DIGITAL

Another area that is rapidly becoming more data-intensive is power generation. In the past, power plants of all types (gas, coal, nuclear, hydro, and others) were typically managed at the individual facility level with minimal regard to the larger context.

At this traditional level are the physical operations and health of each generating asset. Here, data related to maintenance and basic performance is collected. Maintenance schedules are typically set on a routine calendar basis, rather than within the context of the actual physical conditions of the plant. A small number of sensors on the plant trigger alarms, which occur frequently enough that they are often disregarded, making it difficult to discern the true signals from the background noise.

The data provided at this operational level is often not contextualized or analyzed, so that unexpected outages can occur, leading to exceedingly costly downtime (between 3% and 8% of annual availability) and expensive emergency repairs. In addition, spare parts inventories need to be high in order to cope with these uncertainties. At the same time the plants operate with relatively little knowledge of the outside world, including potential market opportunities.[19]

But what if we could change all that? What if we could create real-time awareness of that operating environment to run those machines more efficiently? As it turns out, we can. With the rapidly expanding Industrial Internet of Things (IIOT), some of the major power generation and infrastructure companies are changing the way their generation fleets operate.[20] There are typically three levels of activity where this metamorphosis occurs.[21]

The first is at the generating plant level, where a host of new sensors are installed in the context of an operating dashboard and an IT system that provides real intelligence. GE and some of its competitors have made a big bet on this rapidly digitizing world. By collecting important data points, applying the right algorithms, and improving their insight into how the machines are actually working, these companies can both reduce costs and access new value streams.

In GE's case, it created "digital twins," that is, virtual IT facsimiles of the actual physical assets to offer generation plant operators improved visibility as to what is happening in the plant. This approach is akin to creating an ongoing industrial MRI (magnetic resonance imaging), so one has constant situational awareness with respect to the power plant. A typical decade-old power plant might have 3,000 sensors providing real-time information to operators, while a new plant may have 5,000–6,000 points. The trick is to identify what information is important to measure, how to combine the critical data elements, and what algorithms to apply to the data. In other words, you need the correct raw information to which you apply the right questions. That combination of data and questions yields valuable operational insights.

Properly interpreted early signals can foreshadow problems that could occur in the future. For example, a minute vibration or change in the internal heat of the turbine can be detected earlier, and the machine taken down for predictive maintenance at a time of the operator's choosing. Unexpected outages decrease dramatically, and maintenance can be scheduled based on actual plant conditions. Fewer false alarms occur, and operators know when they have to address something serious. Furthermore, the history of the plant can be interpreted from the data, rather than from the memory of the workforce.

Mastery at the plant operating technology (OT) level can cut operating expenses, minimize unexpected outages, and offer a safer environment, but still does not result in optimized economic performance because it doesn't incorporate key external variables (the IT variables)—such as weather forecasts, and what the remainder of the generating fleet is doing.

The second step in that equation is to involve the power traders who have a view to the markets. If power traders have better insights into how plants can actually operate, they can make better business decisions. Unfortunately, they often don't. Plant managers tend to operate conservatively, given a choice, and may not be fully aware of the economic opportunities. At the same time, traders may not be fully aware of the plant's actual limitations and the stresses that can result from their requests.

For example, optimized participation in a day-ahead market with high prices could result in a significant premium *if* specified plants can safely operate the following day at (or beyond) their full rated nameplate capacity. In the past, it was not always easy to know with any degree of confidence what the specific plant limitations were. However, with an accurate real-time view into the parameters of the power plant, operators and fleet managers can be confident that they can overfire their machines when situations dictate. Better data—if intelligently marshaled—allows for the creation of "what if" scenarios that result in more operational flexibility.

The third step is to combine the individual plant operating data with the wholesale market intelligence, and assess that information at the fleet level to optimize performance of *all* of the assets across the generation fleet. This must be done within the context of the market opportunities and each individual asset's capabilities, physical condition, operating costs, maintenance schedules, and other relevant factors. Doing so will result in an orchestrated approach through which all of the assets are optimized. The best strategic and economic outcomes are obtained when all of the operational and economic trade-offs are taken into account.[22] When that infrequent $9,000-per-hour day comes along in Texas, you want to know exactly how far you can push your fleet, for how long, and what trade-offs you are making in the process.

It's not just the conventional fossil-fired assets that benefit from an enhanced digital approach. Large wind farms and utility-scale solar arrays are seeing their own gains, both from optimization during the siting process and from optimized operations. It's worth noting, though, that as the grid becomes increasingly digitized at all levels, cybersecurity becomes an increasingly critical issue. Any cyberthreat to the power grid represents a potentially existential challenge. The growing interconnection and digitalization of the disparate elements in the evolving smart grid can increase that threat dramatically if proper care is not taken.

6

FROM CON EDISON TO MY EDISON

It has too often been observed that the grid of the early 2000s would have been easily recognizable to Thomas Edison, since so little had changed. That truism is about to be tossed out the window, if it hasn't been already. These days, a good deal is changing on the grid, and one area seeing rapid evolution is the residential sector, where many individuals now have the ability to own their own little power plants.

In mid-June 2020 a friend of a friend of mine, Peter Frisch, emailed me out of the blue. He asked if I could help him think through a possible solar purchase for his home on the North Shore of Massachusetts. It occurred to me that his experience of going through the decision-making process of whether or not to buy solar panels for his roof could be an enlightening one, so I agreed to get involved.

APPLES TO APPLES

Frisch is a partner in a wealth management firm so he knows his way around money, which was helpful in our conversation, as there were a number of things that I would not need to explain to him. For example, I asked whether he intended to finance the panels himself or go through a third party with a lease or purchase power agreement—the latter involving a fixed payment per kilowatt-hour of energy generated, often with an annual price escalator.[1] Frisch's response was immediate: he'd take on that burden himself, since he felt he could get a better interest rate from a bank versus going through the solar company.[2]

Frisch also needed a new roof, which was actually a good thing. Solar companies won't install new panels on a roof in poor condition, for obvious reasons. Armed with that bit of knowledge, I suggested to Frisch he could go one of two routes. First, he could look at a combination solar and new roof deal.

The most well-known of these possibilities is Tesla's somewhat infamous solar roof tile. The concept here is fascinating: Tesla promises to manufacture and install building-integrated photovoltaic (BIPV) roofing shingles on your roof. The panels do double duty, protecting the residence while harvesting the sun on the right sides of your home (north-facing in the northern hemisphere is a no-go), and address a concern of many homeowners who find the typical black or blue solar panels unappealing. For the areas of the roof where the solar gain would be minimal, similar roofing tiles would be used, but they would not have any photovoltaic capability. To the viewer, all Tesla shingles—whether they are tiny power plants or not—look the same.

Tesla's solar shingle concept was rolled out by founder and CEO Elon Musk at a late October 2016 event on a Hollywood lot—the suburban neighborhood set of "Desperate Housewives." Musk had invited a large crowd of reporters specifically to unveil his new solar roof concept. According to one report, partway through his speech, he inquired if anybody had noticed the solar panels yet. "The interesting thing," Musk commented, "is that the houses you see around you are all solar houses. Did you notice?"[3]

Indeed, nobody had discerned that the various slates, Tuscan tiles, and shingled roofs from the houses on the set were meant to be solar panels, and the event created quite an uproar in the industry, as well as no small amount of skepticism. In fact, those particular Desperate Housewife set installations were later revealed to have been nonfunctional.[4]

For the next nearly four years, Tesla struggled to get production going, but in May 2020 it announced that its dedicated factory in Buffalo, New York, was finally capable of manufacturing 4 MW of product per week.[5] In June, it followed that with an announcement that it was refunding $1,000 deposits (some made as far back as 2017) to customers in some states where it would not be offering the product.

I was wary of steering Frisch in that direction at this time (though it might be an entirely different landscape in a few years, and Tesla has shown recent progress), so in our first conversation I suggested he talk to roofers associated with GAF, the large roofing materials company. They offer a combined new roof/solar installation that could be bought and installed in a single integrated transaction. I also asked that he look at EnergySage—a platform that facilitates bid comparison in much the same way one can compete among mortgage companies for the best offer.

Being an individual who does his homework thoroughly, Frisch got back to me the next week and we set up a Zoom call to discuss the various options and see if we could arrive at some clarity. He shared his screen and we got down to work, first pulling up the EnergySage platform.[6] He had previously entered in his address, utility, and energy consumption data, and within a few days, he had four competing bids, all visible on a single web page and expressed in the same format.

The summary information included: 1) a labor warranty (ranging by vendor from 10 to 25 years), a price per watt (a key metric for comparison); 2) the total net investment (after the effect of rebates like the federal solar investment tax credit—ITC—26% as of this writing);[7] 3) the amount one's electricity consumption would be offset (the goal is to get as close to 100% as possible in states offering net energy metering—more on that topic later); 4) estimated total savings over 20 years; 5) the type of panel to be installed and panel warranty; 6) payback—the length of time before one would see their investment repaid; and 7) a loan option.

Each summary was accompanied by a more detailed quote for each installer, again in the same format to allow for easy comparison. Here, one could see the number of panels included in the quote, the projected system size, the number of watts per panel, the number of inverters (the device that converts the DC power to the AC that the grid uses), and the inverter manufacturer.

After we looked at that data and discussed panel quality, inverters, and a number of other issues, Frisch pulled up the quote from the GAF-certified local roofer-cum-solar-installer based in the neighboring town. This quote included an estimate for a new roof, similar to individual roofing quotes Frisch had obtained, as well as an estimate for the solar installation. The pricing was similar on a per-watt basis, but what stuck out was the proposed number of panels: 34, compared to the 20 to 24 that other bidders were suggesting. That was confusing, and certainly raised the overall cost, to $23,172 after incentives. This compared somewhat unfavorably to other bids in the EnergySage platform that by themselves varied widely—from $14,111 to $21,109, even though they purported to offer fairly similar products and outcomes.

THE IMPORTANCE OF NET METERING

The total number of panels and the expected output is critical to the economics of the project because of a concept called net energy metering, or simply, net metering. Net metering is an incentive approach used in 40 states that essentially allows one to store surplus solar energy in the power grid. Thus, for example, during a summer day, on-site solar panels are producing far more energy than is being consumed in the residence. That flow of surplus electrons is exported to the grid and used by other customers. You get a credit for that surplus, generally equal in value to what you would have paid for each kilowatt-hour (minus a few charges that you still continue to pay), and those get banked to your account.

At night, and during the shorter days of the year, however, those same solar panels produce a lot less electricity, or none at all. That's when you use up the credits and offset the cost of the electricity you consume. In many states, if you overproduce, you may only get a fraction of that value, and in some states, you get nothing at all. Thus, the goal of solar installers is to match output as closely as possible to the annual amount of energy you consume.

So, Frisch and I puzzled through the numbers related to the estimated panels (some of which were of different sizes, ranging from 325 to 370 watts—in fact, only two were of the same size). We looked at the quotes and compared expected output with Frisch's last 12 months of consumption. Getting to the bottom of that discrepancy was going to take a bit of work, which is not a good thing if one is trying to shorten the sales cycle. Then we discussed the merits of putting on a new roof and then adding the solar later, or installing both at the same time.

It was getting pretty complicated, so I decided to make matters worse and throw Frisch the "community solar" curveball (why not see if I could make his head explode?).

COMMUNITY SOLAR

In a number of states around the country, Massachusetts being one of them, electricity consumers can participate in rooftop-type programs without actually having to install any panels.[8] These programs involve a centralized installation of a larger solar array, often between 1 and 5 MW (there are caps on minimum and maximum sizes that vary by state), and then a sharing (usually by percentage) of that whole, with each off-taker/subscriber getting a certain part. In different programs, one can either buy the output upfront, or become a subscriber and get a reduction off of the bill.

I have been a community solar customer since 2014 because of my neighbor's large maple tree shading my house, which would have rendered solar panels useless (That was the same tree that would eventually blow over

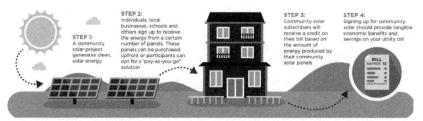

No roof? No problem. *VoteSolar.org,* **https://votesolar.org/files/5115/3183/2131/** CommunitySolarInfographic.png

in a 2019 storm and crush my Prius, which pushed me into my Hyundai lease, so it's had a lot to do with shaping my sustainable energy journey.) My only option for getting solar was community solar.

So in June 2014, I signed a 16-page agreement and for $5,775 I bought a 20-year interest in a solar farm in Rehoboth, a town in southeastern Massachusetts. My five panels' 1,500-watt output is equivalent to .150512% of the total array's production, so each month, I get a credit equal to the entire solar farm output, multiplied by .150512%. That is in turn multiplied by the retail rate I would have paid to my utility to develop the monthly credit on my bill.

A year after my initial commitment, I signed up for the output of another 20 panels. This 20-panel addition allowed me to offset the remainder of my consumption, through an entirely different subscription-based approach that is quite common in the community solar industry. Each month I receive an invoice from my community solar provider for a specific amount. In theory, it's equal to the monthly solar electricity production of the 20 panels multiplied by the utility rate, but I only see a simple amount described as credit. That is the amount that will appear on my monthly utility bill a few weeks later. The amount I am billed on the invoice is equal to that amount minus 15%.

It's a somewhat artificial construct, but it allows far more individuals to engage with distributed solar energy. Not just folks like me with trees, but also apartment dwellers and people with unsuitable roofs. In general, just as with on-site installations, the goal is to offset as much of your annual electricity consumption as possible by estimating the right number of panels and locking in your discount.[9] On an annual basis, I end up building a sizable credit into late summer and early fall, to the tune of over $300. By February, I've burned through most of that, and I typically see one electric bill where I actually owe the utility money (last year I had a single bill for about $100). But by March, the sun is higher in the sky, the panels are firing away, and I begin building credits again to last me through the next dark winter. In fact, with panels angled to the sun, March can be one of my better months, and then April showers typically come along and rain on my solar parade, reducing the monthly output.

Frisch and I ended our Zoom call that day with Frisch pondering numerous options, and probably totally confused, but that can be the nature of the business these days. It can be a bewildering process with a lot of options to choose from.

Confusing as it may be, that hasn't stopped the American consumer from buying solar. California alone has well over 1 million solarized rooftops and the United States surpassed 2 million solar installations in the spring of 2019, with that number having doubled in just three years. A leading analyst forecasted a total of 3 million by the end of 2021 and 4 million just a year later.[10] By 2024, it is expected that the United States may see one solar installation every minute, which is 10 times the rate of 2010.[11]

The entire industry has come a long way since it first really kicked into gear about a decade ago. The cost of installed residential solar has dropped dramatically since 2010, led by the falling costs of modules, but also being aided by declining costs in other aspects of the system including the racking that supports the panels, the inverters that connect them to the grid, and labor costs. Today, a residential solar array that might have cost roughly $8 per watt 10 years ago[12] comes in at under $3.[13]

The speed with which transactions occur is also much faster than it was just a half-decade ago. At that time, the average timeframe for a customer to go solar was nearly nine months.[14] The entire process used to involve a physical site visit with local measurements taken with respect to the solar potential. Today, that process can be done in close to real time in a conversation between vendor and prospect, using satellite photographs and online tools. The COVID-19 crisis further accelerated the move to online sales, with some of the leading companies like SunRun—the country's largest installer—hitting record sales numbers in April during the pandemic.[15]

It's not only the sales process that is vastly faster and more efficient. Companies such as SunRun are also using drone-based site inspections to move to contact-free and rapid inspections. At the same time, a growing number of municipal offices in the country are offering online permitting services. In late April, SunRun proved the art of the possible in San Luis Obispo, California, signing a customer in the morning, receiving the digital permits during the course of the day, and installing a system that same afternoon.[16]

ADDING BATTERIES TO THE EQUATION

The entire residential solar industry is becoming more effective, and now it's moving on to the next thing: adding a healthy dash of batteries into many of the systems being installed.

There are several reasons for this at the residential level (and often for commercial customers as well). The first is that batteries offer the home-owner some peace of mind. It's no accident that Generac, the leading backup generator company, has recently moved into the solar-plus-storage arena. A small number of utilities, such as Vermont's Green Mountain Power (GMP)[17] and New Hampshire's Liberty Utilities,[18] work with cus-tomers to install batteries on their premises for backup power. GMP was the first utility in the country to pioneer such an approach and now has a rate that provides incentives of up to $10,500 per customer for battery installations. At least 100 customers per year can participate, and they can choose among a variety of approved batteries. GMP can also use the bat-teries when they are not needed for backup power to store energy for use during high-priced peak demand periods. GMP has a separate program just for Tesla Powerwall batteries, through which customers either pay $55 per month for two batteries in a 10-year lease (with five additional years at no cost), or an upfront fee of $5,500. They also agree to share their batteries' energy with GMP during peak periods.[19]

Companies like SunRun, Generac, SunPower, Enphase (originally an inverter manufacturer that got into batteries) all see a huge opportunity in the space. That's especially the case where rooftop solar arrays can be used to power on-site batteries, providing customers with longer-term backup power supplies. The amounts provided may not be enough to power an energy hungry air conditioner, but they will be sufficient to support essen-tial services such as lights, refrigerators, and computers.

SunRun has seen a very significant opportunity in northern California after the wildfires and ensuing preemptive power outages by local utility PG&E that affected millions of people. As a result, in the Bay Area over 60% of April solar sales were combined with batteries, and SunRun has installed over 10,000 of its battery systems nationally, growing 50% over the previous year.

Another value that storage provides to the residential consumer is the ability to store electricity generated on the rooftop for use at a later date. This is highly valuable in markets that have customer rates that vary based on the time of day, such as the one employed by San Diego Gas & Electric (SDG&E). Every SDG&E customer with on-site solar panels must be on a time-of-use rate. As one can see from the figure below, there are two super off-peak rates where electricity is relatively cheap: one from midnight to 6:00 a.m. when demand is low and less-expensive generation resources are required, and one from 10:00 a.m. to 2:00 p.m. when the market is saturated with solar power from millions of panels across the service territory. One can also see that the rates during those periods are as low as 19 cents per kWh (if you consume less than 130% of your normal average baseline consumption—welcome to the arcane and inscrutable logic of utility rates, which are intended to reflect costs regardless of how much they confound customers).

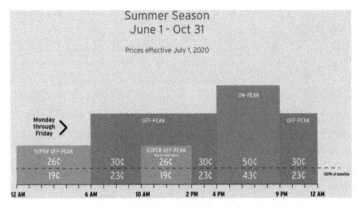

Time-of-use rates are begging for batteries. *San Diego Gas & Electric,* **https://www.sdge.com/whenmatters#plans**

By contrast, the evening on-peak prices from 4:00 p.m. to 9:00 p.m. are 43 cents—well over twice the super off-peak rate. This provides a predictable and potentially lucrative ability to "buy low and sell high," in the sense that one can charge the batteries either from the grid in the wee hours of the night or from rooftop solar panels during the middle of a sunny day and then avoid consuming grid power at 43 cents during the higher priced tariff hours. Day in and day out, that value could add up quickly, and all the while the battery is also providing the customer with a backup energy supply in the event of an outage.

WHEN NET METERING GOES AWAY

Then there's the issue of net metering discussed previously. Many states permit net metering up to a fixed percentage of the utility's total energy sales (often in the range of 5%–10% of total utility sales). Once those thresholds are reached, net metering is replaced with other structures, but they generally result in a much lower rate of compensation for owners of on-site solar. Depending on the difference in the rate one pays the utility and the amount the utility pays to the owner for excess solar production, there is often a profound economic incentive to add storage if net metering is eliminated.

ADDITIONAL VALUE STREAMS

There is another source of value that is growing rapidly in some areas of the country (and the world) and that it is the ability to aggregate and share those distributed resources with the grid, acting as a "virtual power plant." As discussed in chapter 10 on energy storage, these value streams can be quite significant, and they can be derived from the sale of different "products."

For example, batteries can be aggregated and utilized to provide fast-reacting stability at the local distribution utility level, responding in a fraction of a second. They may provide a similar service to the larger wholesale grid. Or they may provide something called "capacity"—the ability to provide a specified level of megawatts for as long as four hours. A number of vendors are now aggregating and delivering value from these distributed storage resources to various buyers. SunRun, for example, has contracts with the New England grid to offer 20 MW of aggregated battery capacity in 2022 from 5,000 homes for up to four hours.[20] The company signed a contract in July 2020 with three of California's energy suppliers to aggregate 6,000 home battery systems and provide backup power to vulnerable customers in the event of electricity shutoffs during wildfire season.[21] Meanwhile, competitor SunPower inked its own, similar deal with the New England grid operator to sell 11 MW of capacity services by 2023.[22] This has the potential to be big business: By summer of 2020, SunRun claimed to have over $50 million in revenue from such grid services either booked or in an advanced pipeline.

Most of these projects are still being developed in states where strong incentives for solar energy and batteries propel markets forward. However, a Utah-based project announced last year by German solar and storage company sonnen Inc. (now a subsidiary of Shell) suggests that the potential will soon be more widespread. In that undertaking, sonnen teamed up with developer The Wasatch Group and utility Rocky Mountain Power to create a virtual power plant featuring solar panels at an apartment community outside Salt Lake City. That $34 million project includes 5.2 MW of rooftop solar for the entire 22-building complex, as well as 600 individual batteries situated in apartments that offer backup power to residents as well as 12.6 MWh of energy storage for use by Rocky Mountain Power. The utility committed to paying $3.27 million for the use of the aggregated storage resource.[23]

Can you find the battery? *sonnen USA,* **https://sonnenusa.com/en/wasatch-group-sonnen-and-rocky-mountain-power-launch-first-kind-all-electric-apartment-community/**

GOODBYE, UTILITY?

Sonnen is working on similar projects in other parts of the world, including Europe and Australia. One sonnen project in particular should strike fear in the hearts of utility executives. In Australia, the company had teamed up with home developer and solar installer Natural Solar on a 3,000-home residential development combining solar panels and batteries. Under its program, customers make an upfront investment in panels and batteries and then pay an ongoing monthly fee.

In September 2019 I spoke to Stephen Fenech, Natural Solar's first customer, who paid $19,000 AUD (approximately $13,000 USD) for a 20-year deal. In exchange, Natural Solar installed 27 panels rated at 280 watts (a total of about 7.6 kW) on his roof and a 10 kWh sonnen battery in his laundry room.[24] Fenech pays an additional $40 (AUS) monthly fee under a program called "sonnenFlat." He estimated he was saving enough to achieve a return on his investment in five and a half years, between the savings on powering his Tesla and what he would have paid on his electric bill.

Sonnen offers three different plans based on the size of the home, and uses a sophisticated platform that creates a virtual power plant and allows sonnen to sell various services to the grid. The software functions similarly to the platforms of other vendors such as SunPower and SunRun, or distributed battery company Stem, managing critical internal battery parameters such as temperature and state of charge while constantly observing the market and grid conditions for opportunities to deploy the fleet of batteries. Algorithms tell the batteries when to absorb energy from the grid and charge up, or when to inject power back into the grid, optimizing the overall economic profile. The ability to harvest these revenue streams creates a virtuous cycle for sonnen, allowing the company to offer its products to the customer at lower prices.

For his part, Fenech enjoys the peace of mind and the ability to manage his utility bill within the context of his household budget. As far as Fenech is concerned, seasonal variation in usage—and the need to budget for it—is not his problem anymore. In fact, it has become difficult to determine what he would have actually paid on his electric bill, since he hasn't seen one since the day he signed his contract. As far as Fenech is concerned, the utility really doesn't exist. These days, sonnen gets that utility bill and pays it, without him ever having to think about it. That should be cause for concern among utility execs across the planet.

I caught up with Chris Williams, the Australian CEO of Natural Solar, via Zoom in mid-August 2020, to get a sense of how the solar-plus-battery industry was proceeding. He was excited to note that his company was working on a number of residential and master-planned developments, and Natural Solar was on the cusp of signing a deal with the country's largest developer of new homes—about 6,000 annually. Williams expected that 2,000 to 3,000 buyers each year would opt into the sonnenFlat product. "Virtual power plants are popping up across the country," he stated. As a consequence, solar providers were no longer just solar installers anymore—they had morphed into "trying to be an energy supplier. It's changing the discussion from solar and storage to an energy offering."

Williams was also excited about a program in which the New South Wales government is offering a $14,000 AUS (about $10,000 USD) interest-free loan for solar and batteries. "If a customer is paying $150 on their monthly bill, we can add solar and a battery. They pay nothing upfront and receive a guaranteed discount of 20%–40% We are not discussing a solar and battery," he emphasized, "We are discussing an energy plan. Households may not have been as engaged before, but now they see a lower electric bill, and you get mass-market adoption. Households that couldn't previously afford it are now getting a return straight away."

As the components become increasingly cheaper, this dynamic will soon begin to run on steroids. Batteries will increasingly flood the landscape, to the point where SunRun's CEO Lynn Jurich says, "It is clear that solar plus storage will be the standard offering in the coming years."[25]

GRID DEFECTION VERSUS LOAD DEFECTION

In the case of sonnen's Australian customers, or those of SunRun, SunPower, and a host of other companies in the United States, most utilities are not concerned that their customers will defect from the grid entirely (a la the cabin in the woods where one cannot have air conditioning or use two 800-watt hair dryers at the same time). The economic argument is that the combined efficiencies of everybody being connected to the same system and using electricity in different ways at different times creates efficiencies that make it more sensible for people to stay connected.

The logic suggests that it would be economically inefficient—and costly to the individual consumer—to size each home with a battery that would be capable of meeting maximum peak demand. Nobody wants to limit how and when power can be consumed in order to be independent of the grid. To date, then, the concern has not been about consumers leaving the grid, that is, "grid defection," but more about having a large percentage of their energy being produced and stored by solar and batteries, that is, "load defection." After all, if a growing number of customers only take 20% of their historic usage from the grid, that will put a serious crimp in utility revenues.

But what might happen if solar panels and inverters continue to get both more efficient and less expensive? A more efficient panel means you can get more energy from the same space. Less-expensive panels (and racks and inverters) means you can buy more of them. Perhaps more critically, what might occur if battery manufacturers continue to deliver on their promises of cheaper, safer, more durable, and more powerful batteries? What are the potential implications if a company like Contemporary Amperex Technology—one of the world's largest battery manufacturers—really can bring to the market a battery that offers 1.2 million miles of distance to a car over its lifetime, compared with today's batteries that offer only about one-eighth of that distance, at only a 10% premium?[26] One could charge and discharge that battery at will, without regard to potential limitations on its cycle life. And as the entire ecosystem of batteries and solar panels becomes increasingly more cost-effective, perhaps there does arrive a day when a consumer could buy a large and durable battery system for the home and ditch the grid entirely.

THE NEXT INEVITABLE MASH-UP:
VEHICLE TO SOLAR-POWERED HOME

Perhaps more likely, though—and probably inevitable—is a scenario that is only now coming into view, and that is vehicle-to-grid, or V2G. V2G technology would allow you to both charge your vehicle at the most desired time of the day as well as to drain energy from the car back to the grid. Today, the largest EV is Tesla's Model S 100D, and it will get you 400 miles down the road with every charge. Its 100-kWh battery also holds enough energy to supply the average American household with over three days' worth of electricity. By comparison, most battery storage systems offered for home use are much smaller: SunRun's BrightBox is 9.3 kWh,[27] while Generac's current basic system is 8.6 kWh, scalable to 17.1 kWh,[28] and Tesla's Powerwall is 13.5 kWh.[29]

Today, most of the millions of electric vehicles on the road do not offer V2G services, despite having much larger batteries than residential storage systems (the 2020 Hyundai Ioniq with its 170 miles of range has a 38.3 kWh battery).[30] Most EV manufactures are wary of having their vehicles pull double duty and offer services to the grid, mostly due to warranty concerns. That's because most batteries are generally only rated for between 1,000 and 2,500 cycles. So, for example, if you used the car battery every day in San Diego to play with daily time-of-use rates, the battery would last between three and seven years.

However, that landscape is now changing. Nissan in particular has begun to embrace the V2G promise. In 2018, it unveiled a plan so that "owners of Nissan's electric vehicles will be able to easily connect their cars with energy systems to charge their batteries, power homes and businesses or feed energy back to power grids."[31] Nissan has had vehicle-to-home capability since 2012 in Japan, and although it hasn't announced that capability in the United States yet, third parties have announced plans to offer technology that will allow Nissan Leafs to provide power directly to homes (for about $4,000).[32] In August 2020, Nissan went a step further in Japan, allowing drivers to pay for parking at its Nissan Pavilion in Yokohama with the electricity from their vehicles.[33]

A car battery can now be used to pay for parking. EV batteries can in theory supply power to a home for one to three days, depending on the size of the battery. And the batteries themselves are about to see massive gains in cycle life. A vehicle-to-home scenario for backup power, and for interaction with the power grid, is not only possible—it is likely inevitable.

7

THE SUN ALSO RISES

On a mid-August morning, I decided to pay a visit to some solar panels that sit among thousands in a Massachusetts field just east of the border with Rhode Island. Six years ago, I bought the output of five panels' worth of electricity for $5,775 in a community solar project that would credit my electric bill for the energy produced by those panels.[1]

My contract specified that I would actually receive serial numbers for my specific panels, even though I was to be credited with a percentage of the project's total hourly output. I'm not sure I ever did receive the paperwork with the serial numbers, but nonetheless, I wanted to make the pilgrimage and actually *see* the place where my panels were generating electricity and dumping it into the grid.

I contacted the owner of the facility, who kindly agreed to arrange a visit. So I jumped into my Hyundai Ioniq EV with its 124 rated miles of range on a 132-mile round trip. I wanted to see if I could make it without charging the battery (but threw the charging cord in the car, just in case). The indicator told me I had 135 miles of range, which meant I would be cutting it close. So the first thing I did when I climbed into the car was set the air conditioner up to 75 degrees. That would minimize the amount of energy the battery would need to cool the car, and the range indicator immediately jumped to 165 miles.[2]

I pulled into a dusty field 75 minutes later and met the facility owner's asset manager. We exchanged virtual handshakes, bowing to the realities of our COVID-19 world. He unlocked the gate, and we walked the length of the property past row after row of dark blue panels. Each row included a galvanized support structure that raised the panels off the ground and tilted them toward the sun at a fixed angle of 20 degrees with an orientation (azimuth) directly south at 180 degrees. Each structure also had a gray

Five of those are mine. *Google Earth*

electrical box attached and a conduit into the ground that directed all of the electricity to a 750 kW transformer. Underneath the rack and below each panel were the junction boxes that connected them together and—almost hidden by the rack itself and nearly impossible to see—the magic serial numbers. Looking for mine would have been needle-in-the-haystack stuff.

There was not that much to see, really, as I had suspected there wouldn't be. The real magic is what happens inside those panels. Being responsible for operations, the asset manager oversees the company's various projects to ensure they are functioning as expected. With decades of experience working with all types of projects including wind and conventional power plants, he made a comment as we traipsed down the rows that I found intriguing: Gas-fired plants often run for many hundreds of hours, and nuclear plants for well over a year before being cycled off or taken down for maintenance. By contrast, "solar arrays cycle completely every single day" over their expected 20- or 30-year life spans. That's a tough operating regime, but it's no accident. They've been designed and built to do just that.

Close-up of a solar panel: turning photons into electrons. *Photo courtesy of the author*

RUNNING THE TESTING GAUNTLET

Imagine you are a solar module coming to the United States and happily anticipating two or three productive decades out in the field.[3] You hope to join the tens of millions of panels in the field already generating 85,000 MW of power as of mid-2020, enough to supply almost 16 million homes.[4] You started from silicon ingots cut into cells that were then polished, equipped with busbars (little wires to move the energy you are collecting from the sun), and assembled within a metal frame. Machines then glued a glass panel on your front and a backsheet to your backside (although if you are bifacial, you are transparent on both sides). Somebody also added a small junction box to keep the power generated moving in one direction and connect you to the next panel. Then you were stacked onto a forklift, dropped into a container, and shipped to the United States.[5]

Then you have the utter misfortune of meeting Jenya Meydbray, CEO of PV Evolution Labs (PVEL), whose team is going to lock you in a 12 cubic feet sealed space and accelerate your aging process. Meydbray is going to find out how you will stand up to decades of life out in the field. So you'll be heated to 85 degrees C (185 degrees F) and then cooled to −40 degrees C (−40 degrees F), seven times a day, as many as 800 times in total to see how your glues and solder joints hold up after many tens of thousands of hours on the job, what happens to your cells and integrated plastics, and how much energy you'll be capable of generating over your lifetime. At the end of the day, this is about finding out whether you and your brethren are truly cut out for the task, how often you can cycle, how productive you will be, and for how long. After all, a relatively small percentage difference in output can have a profound impact on the balance sheet over a few decades, and many millions of investors' dollars can hang in the balance.

AN EDUCATED AND
EXPERIENCED VIEW OF THE INDUSTRY

I've contacted Meydbray occasionally over the years and interviewed him for various pieces on the solar industry, and he's just the person to explain the ongoing evolution of solar photovoltaic (PV) technology. So, I scheduled a Zoom chat with him in mid-June to discuss the current and future trajectory of solar technology, and the likely developments to come. At the time of our conversation, most of his team members were still on COVID-19 lockdown, working from home, with a skeleton staff in the lab doing the actual physical work.

He described the lab as a 12,000 square foot, heavily disinfected, clean space closed to the public, with the workers wearing masks and practicing social distancing. The rest of the team work remotely "with kids in the background," but other than that still functioning reasonably well. By mid-winter of 2020, Meydbray's group was already seeing the economic shock-waves of COVID-19 rippling out from China well before most Americans did. PVEL's business somewhat tracked the Chinese experience, with a slowdown in late winter as the Chinese economy largely shut down, but the company was almost back to normal business levels by March.

HOW PANELS FAIL

Meydbray described the basic elements of a solar panel as follows: "Inside a solar panel is a bunch of individual solar cells—72 or 144 or 156 individual pieces. They are little silicon wafers processed and wired together with some flat or round wire soldered to the front of one cell and the back of the next, creating an internal circuit, and that is glued to glass using an encapsulant."[6] On the back side of the panel, there's either more glass (if it's a bifacial panel that can harvest sunlight from both sides of the panel) or an opaque backsheet. There's also a metal frame around the perimeter of the panel for handling, mounting, and shipping. The internal wiring goes through a hole in the back, and connects with a junction box that connects each panel to the next.

However, that internal circuit must maintain its continuity and its integrity in harsh outdoor environments, and over time, with exposure to the sun and temperature variations, solar modules inevitably begin to degrade. Meydbray ticked off a list of potential failures: "The cells themselves will degrade. The package has to stay glued together. The junction box on the back needs to stay glued to the back—you don't want that to fall off. Then the wires have insulation and plastic connections, and they need to stay seated so they can't get loose. And you can't let moisture get in." Nonetheless, and despite best efforts and technological improvements, solar modules inevitably degrade. The critical questions are simply, "how fast, and in what fashion?"

It's Meydbray's job to figure out the ways a solar panel will age in the field. A panel's enemies are numerous. For example, there's the critical thermal cycle. During the day it gets hot (and many panels sit on dark rooftops—further enhancing that dynamic), so the panels expand. At night, the ambient temperatures cool down, causing thermal contraction. Since the

Many ways to fail. *Photo courtesy of PVEL*

module elements are made of different materials—silicon, glass, metals, and plastics, much of it held together by glues—they expand and contract at different rates. This is a process known in the industry as thermo-mechanical fatigue (power plants are vulnerable to this too—and power plant life spans are defined by cycles as well).

In addition to the thermal cycles, outdoor humidity can lead to pieces becoming unglued. Then there's ultraviolet (UV) sunlight that results in degradation of plastics and polymers. "If you've seen a kid's plastic toy in the yard all summer," Meydbray said, "the same thing happens in a solar panel. Those plastics have to be robust to withstand UV light for three decades, which is a very long time."

I asked Meydbray what would constitute a good degradation rate. "One half a percent per year. Everybody would be happy with that," he said, "all the investors would, and that's typically the number in the spreadsheet." Premium (and somewhat more expensive) panels might only degrade by .20% to .25% annually, while some poor performers that he's tested have degraded by 20% over just the first few years. "Well-built panels will perform almost as well" year after year, "and really crappy ones won't perform at all. Medium-crappy ones will degrade at a 6%, 7%, 8%, or 20% rate. You might see things like cracks in solder joints, and the crack propagates."

A more difficult challenge in testing is the simple fact that the technology is continuing to evolve. Just as wind turbines are getting more powerful, and battery chemistries continue to constantly improve. "Solar tech is evolving pretty quickly, so the aging behaviors vary," Meydbray notes. A new panel may look the same as one from a few years ago, "but inside things are now different." There may be new polymers with different chemistries, for example, or the cells themselves may be thinner.

All of that means that Meydbray is likely to have a job for years to come. He related a story from 2009, when he was raising money to build the lab. "There was some recurring feedback that I got, with people saying, 'Once you're done aren't you done, and the business isn't needed anymore?' My response was, 'When Intel designs the next chip, do they fire the R&D department?' You are never done."

A TECHNOLOGICAL ARMS RACE

From a technology standpoint, the design and manufacture of solar panels is still a relatively immature industry, so there is plenty of room for steady incremental improvements. As it turns out, the pace of these improvements has been relatively predictable. Meydbray commented that he had been seeing conversion efficiencies—the rate at which panels convert sunlight into electricity—increasing at roughly half a percent per year for more than a decade. That doesn't appear to be slowing down anytime soon. "If anything," he emphasized, "we are actually seeing it accelerate. Innovation has accelerated, and each manufacturer is trying to one-up others in press releases right now."

Among the main drivers of change, he said, is the fact that the modules are getting larger. For years, panel sizes clustered in the 300–350 watt range. But now, he said, we are seeing 400- and even 500-watt panels. One element driving that dynamic is the fact that the industry is making larger wafers that are the basis for the cells. There's a Chinese factory–driven arms race for wafer size, with some industry giants evolving from 166-mm wafers to 180 mm, while others are eyeing 210 mm. Larger wafers ultimately result in larger panels, which can create potential economies of scale and efficiencies across the entire ecosystem.

At the same time, Meydbray cautioned, "there are implications across the board with solar design and plant design. It remains to be determined if one wafer size will prove to be the winner and all eventually consolidate around that, or if there's one for rooftop and one for utility scale—it's kind of up in the air." There are also new trends for how manufacturers set up the wiring inside the modules to reduce the amount of wasted space between each cell in the module, and approaches such as shingling, in which cells slightly overlap so there are no "dead" areas in the panel that do not harvest sunlight.

Despite all the recent progress, the solar industry is currently limited by the tyranny of the prosaic industrial shipping container, which is either 20 or 40 feet long, but always eight feet wide. Meydbray commented that at one point, the solar industry attempted to migrate from a 72-cell panel to a 96-cell panel, which increased the overall width by one-third. However, that restricted the ability to double-stack panels in shipping containers, so that initiative was quickly abandoned. As a consequence, modules can continue to get taller, but they cannot get wider without significantly increasing shipping costs.

The tallest panels Meydbray has seen moving through his lab are fully eight feet. At that point, the question then becomes how efficiently these can be handled by workers in the field during the installation process. "Nobody yet knows where the sweet spot is for the right size for construction workers moving most efficiently," he comments. "Is eight feet too big?

I don't know, and if you talk to construction companies they also don't know." And what are the associated impacts on the rest of the ecosystem, such as the racks in the tracking systems that move these panels to follow the sun? Taller panels become exposed to more potential wind load, and higher torsional twisting forces on the infrastructure may put more stresses on the motors that rotate the panels, thus allowing fewer panels in a single row. These are all issues that are continually being worked out, and Meydbray said, "It's still unclear where the sweet spot is. Nobody really knows what the right number is."

Part of the challenge, he noted, is that the global-scale panel manufacturers are focused solely on selling watts as efficiently as they can. Meanwhile, the developer is burdened with integrating all the pieces. In recent years this issue has been further exacerbated, as many companies that were originally hybrid developers/manufacturers sold off their manufacturing segments.[7] This dynamic of dis-integrated specialization is not unique to the solar industry, and Meydbray aptly summed it up in a metaphor: "There aren't too many car companies that make their own tires."

WHAT MIGHT A PANEL
LOOK LIKE TEN YEARS FROM NOW?

Panel efficiencies have seen a steady march over the years, and as noted, various improvements have resulted in consistent annual gains of approximately .5%.[8] However, silicon-based panels eventually run into the unyielding physical conversion barrier of approximately 33% for a simple silicon solar cell—the Shockley-Queisser efficiency limit.[9] In other words, a simple solar cell can convert roughly one-third of the energy from direct sunlight into electricity, or photons into electrons. Today, the most efficient panels in the market—from SunPower—offer a nearly 23% conversion efficiency (compared with the majority of panels out there, which range between 15% and 20%).[10]

The practical conversion ceiling is around 27%, said Meydbray, which would be about 270 watts per square meter. Thus, with a 2 square meter panel, "we are talking close to a 600-watt solar panel kind of ceiling with today's tech. That probably gets us out five to seven years from now, marching up the efficiency curve in cell technology evolution."

At that point, Meydbray asserted, you need "a step function to the next thing. . . . Then you need to figure out nonsilicon solutions if you want to keep going up." One concept being explored is the utilization of a silicon solar cell as the base, with another cell layered on top, working in tandem. The first cell permits some light to pass through and hit the solar cell on the bottom. "So you get power from the top cell and the bottom cell, and you can keep going up the curve, and there's no reason you cannot make a 40% or 50% solar cell, though we are far from knowing how to do it today." At least not commercially.

In fact, using a similar approach, scientists from the National Renewable Energy Laboratory announced in early 2020 that they had created a multijunction cell with six layers (junctions) to create a world-record conversion efficiency of 39.2%.[11] Each junction harvests energy from a different spectrum.

It's one thing to do that in a lab, and quite another to have a commercially viable panel in the marketplace with the desired traits. However, Meydbray speculates that by the 2030s, we may see an 800- or 900-watt panel: "We are on a trajectory to go there." These panels may involve a class of materials called perovskites that promise both high efficiencies and low costs. Meydbray asserted that every major panel manufacturer is doing perovskite research, but cautioned that perovskites are salts that dissolve in moisture, require a real hermetic seal, and are very unstable. "Today, you are lucky to have it last a week. There's still a long R&D road map, and whether perovskite is the top cell is still to be determined, but there is a pathway to go past the 27% limit and to keep going up." Meydbray speculated that at some future date, "Maybe we will be getting to a point where solar is so cheap, it's not sufficiently compelling to keep doing the research to go up the curve, and the focus will be on costs instead."

That future is uncertain, but what is abundantly clear is that the solar industry is aggressively focused on innovation, and solar's growth has been far beyond Meydbray's initial expectations: "It's a hell of a lot bigger than any of us thought in 2006 when I first got into solar. We were perfectly content selling to pot growers and hobby niche applications, but I guess now it's come much further than that."

TOURING A FACTORY FLOOR

Several years ago I had an opportunity to visit an unnamed assembly facility and tour a factory floor to see how these solar modules were made. Upon entry to the factory floor, the first thing visible were process flow diagrams outlining the sequence of activities and how the parts would be assembled into a finished product. Two identical and parallel assembly lines were set up in this large industrial shed, supporting a finely choreographed interaction between humans and machines. The pace of machinery and people was fast, but also deliberate, with humans located as the connective tissue between many different machines and specific automated processes.

The humans were largely there to ensure quality at the various transition points, as the constituent elements of the panel were soldered, glued, or joined under pressure by sophisticated machines. At one point, where lines of joined cells were being positioned, my guide hit a switch, stopped the process, and extracted a small, thin fractured portion of a cell that had splintered earlier in the process. Even finely instrumented machines make mistakes sometimes, so today there is still a need for people to ensure process integrity.

A critical element of successful and economic plant operation is to minimize these exceptions and related interruptions, so instilling a culture of self-improvement is important. Critical key performance indicators are constantly tracked and root causes identified. Solutions are proposed, fixes and adjustments are made, and the factory continues pushing out panels—24/7, five days a week—into a world increasingly hungry for low-cost solar energy.

It was not hard to imagine the thousands of person-hours that had gone into the design and troubleshooting related to the development of each constituent process, and integration into the whole. It also wasn't hard to imagine that in the near future, the quality control process would largely be outsourced to machines controlled by artificial intelligence and the factory would operate with almost no people.

This was a relatively small module assembly facility, churning out panels at an annual rate of 200 MW of panel capacity, a number that barely qualifies as a drop in the global bucket. China's Jinko Solar, for example, boasts 16,000 MW of annual production capacity, and has delivered over 52,000 MW of solar modules over the company's life span.[12] Global solar installations in 2019 totaled 115,000 MW.[13]

If it wants to survive in that brutally efficient world, the factory I visited will be under constant pressure to cut costs and improve efficiencies, working on extremely tight margins. That's hard news for the factory, but good news for the planet, for developers, and for the consumer.

That dynamic of constantly improving efficiencies in the cells and the panels, combined with ruthlessly efficient supply chains, has resulted in a 90% decline in the installed costs of utility-scale solar energy projects.[14] And that trend is not going to end anytime soon.

SYSTEMS ARE ALSO GETTING
SMARTER AND MORE SOPHISTICATED

It's not just the panels, though. The rest of the costs in the solar ecosystem are declining as well, aided by a healthy dose of IT. In uncomplicated solar installations, such as one sees on most rooftops, solar energy is a pretty simple thing. The panels are mounted on a rack at a fixed angle—between 30 and 45 degrees—and generally facing south (in the northern hemisphere), and they harvest energy from the moment the sun appears until sunset.[15] Ignoring clouds for the moment, they have a predictable output curve during the course of the day, across the entire season.

However, in a utility-scale solar plant things are often quite different, and far more sophisticated. These power plants are often scaled at hundreds of megawatts of capacity. In fact, the largest American PV solar project to date, Solar Star (located just outside of Los Angeles), is rated at 579 MW and covers 3,200 acres (5 square miles).[16] That kind of maximum power output is similar to that of a mid-sized gas-fired generating plant. Solar Star uses 1.7 million panels, all of which are mounted on single-axis trackers that allow the panels to tilt to optimize output.[17]

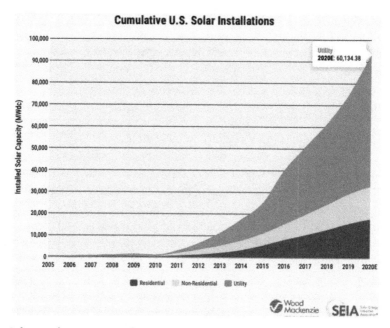

Solar on the way up. *Solar Energy Industries Association,* https://www.seia.org/research-resources/solar-market-insight-report-2020-q2

The entire solar industry has been growing at a rapid clip, with the utility-scale sector increasing fastest of all and taking significant market share in the power generation space. In fact, 40% of the new generation capacity added in the United States in 2019 came from solar power plants.

TRACKING THE SUN

Trackers are increasingly the way to go for larger projects these days (over two-thirds of new project utilize trackers).[18] Adding a motor and intelligent software that infuses panels with the ability to follow the sun from east to west over the course of the day significantly improves project economics.[19]

One company that has been playing in this tracker space for a long time (since 2013) is Nextracker.[20] I had previously interviewed its CEO, Dan Shugar, in mid-2019 when his company had just surpassed the 20,000 MW milestone. As of September 2020, Nextracker had increased that number to well over 40,000 MW, working with developers and owners to install, operate, and maintain tracker-integrated systems around the world.

I had been looking forward to catching up with Shugar in mid-July at the MidWest Solar Expo, where we were both slated to speak. However, by late spring that event had been moved to a "fully immersive 3D virtual platform," so I didn't get the chance to physically interact with Shugar. As it turned out, the closest I got was when his virtual avatar came rushing past mine on his way to the virtual conference hall. I could hear him speaking as he moved by me, and could see his name and title over his head as his voice receded into the distance.[21]

I was nonetheless able to follow up with a videoconference invitation and caught up with Shugar one afternoon as he piloted his Tesla through the California landscape. As we spoke, I lost my connection to him for two minutes as he passed through a tunnel, but we reconnected and continued our conversation as he was recharging his EV at a Tesla charging station. In some ways, Shugar himself is an avatar for the future, and though at times his signal may have been weak, the story he told was loud and clear.

He kicked off the conversation with two simple statistics: when he started work at northern California's Pacific Gas & Electric in the late 1980s, it was the largest investor-owned utility in the country. It had taken the utility over a century to build its system up to 20,000 MW (20 GW) of generating capacity. "Here at Nextracker," he commented, "we were founded with less than $5 million in 2013. And we've got well over 30 gigawatts done in seven years. That kind of sums up what the potential is here. I couldn't have guessed that in my wildest imagination." Shugar recalled a moment back in 1988 when he was looking out at a landscape populated by wind turbines in California's Altamont Pass. "You could see turbines in all directions and it was about 500 MW." He remembers thinking his lifetime goal was to install 500 MW of renewables. It took him about 15 years to reach that goal, he said, and then laughed, "Now we are doing 500 MW a week, and I'm sure the numbers will get bigger."

Two or three decades ago, the industry focus was on solar thermal (hot water, which was far cheaper than PV systems per energy produced), and that focus then migrated to residential rooftop solar. The latter is still making strides, but today Shugar characterizes it as "distant secondary afterthought in terms of actual contribution in terms of the grid and energy. The utility-scale stuff and ground-mounted capacity is 75% of the capacity and energy being installed, versus 25% back then." Shugar also pointed out that the productivity of a utility-scale, ground-based system is about 50% more than on-site solar. "That's because it uses trackers that follow the sun, the panels are professionally maintained, and they are not shadowed."

Of course the hardware—which include the racks and the motors, and the inverters that connect the DC panels to the AC grid—is a critical part of the overall system. But the secret sauce is the software, which his team has developed and is continuously refining. The hardware and software go hand in hand via industrial-grade Zigbee wireless protocol, and a data hub architecture that employs bidirectional communication and collects 4 terabytes of data every five minutes. The first element is the controller, containing the algorithm that "talks" to the panels. Nextracker also has a software system called TrueCapture that enhances overall energy yield by about 4%. Among other things, it ensures that the panels don't shade each other and reduce overall generating output, while also taking topography into account, so that placement of each rack of panels can be optimized when dealing with rolling terrain.

Nextracker also has a complementary package called NX Navigator, and among other things its job is to protect the panels from weather damage. Shugar observed that the company is seeing an increased incidence of weather extremes in recent years—"more hail, crazy rain, and hurricanes." Hail may be the most dreaded word in the industry. He recalled an incident in the spring of 2019 when a 160 MW solar project in West Texas was hit by hail and suffered a $75 million insurance claim.[22]

When there is a threat of hail, operators in the Nextracker 24/7 control room are constantly on the alert for high-priority Doppler radar alerts.

In a hail event, operators can rotate the entire plant to 60 degrees in 90 seconds, "so when hail falls, it glances off the panels, eliminating breakage."

The solar modules aren't the only thing evolving, Shugar indicated. The associated software systems are evolving as well, and now the relevant question becomes, "How do you use machine learning and AI monitoring and telemetry to be able to upgrade these plants to get smarter as they adapt to a changing environment?" Nextracker's approach is to embed a lot of capability in the hardware, and a lot of intelligence in the software and firmware to "try and future-proof the PV systems. Just like the Tesla I'm driving, the firmware keeps getting better." Shugar's company employs a variety of tools including distributed sensors in the hardware. In addition, he said, "We have inclinometers, we measure motor current, battery voltage and temperature, and distribute all that into the system. . . . In the case of the best projects, we're delivering another 20%."

Today's systems are designed to maximize output over all types of weather conditions. For example, during a completely overcast day, the TrueCapture software and mesh network instructs the motors in the racks to position all the panels to be close to horizontal. The resulting difference in output between a panel in a "normal" position and one close to horizontal on days when widespread clouds or pollution block the sunlight (a phenomenon known as isotropic scattering) can be in the double digits. Taken together, all of these improvements have conspired in recent years to increase output, drive down costs, and cut required project acreage in half.

THE ADVANTAGE OF BEING TWO-FACED

Then there's the rapid emergence of bifacial solar panels, which create the capability to harvest energy from both sides of the panel. While the concept has been around for well over a decade, it has really begun to make strides in just the past few years. Recent estimates suggest that the bifacial market will grow tenfold in the next half-decade.[23] Shugar commented that "every major manufacturer is migrating a very large percentage of output to bifacial," which further improves the value proposition for trackers versus fixed-tilt panels.

If a panel can harvest energy from each side, though, that raises an entirely new host of challenges. The main issue is simple and complex at the same time: How high should the panels be raised off the ground to maximize the reflectivity, the so-called albedo effect? Not surprisingly, that albedo effect can vary greatly. Fresh snow, for example, can provide a reflectivity gain of 75%–95%, while dry sand can offer reflectivity somewhere in the range of 35%. Grass ranges from 17% to 28%—on the higher side in the winter, when it's dead and light in color.[24] Raising the panels higher off the ground—even just a foot or two—can significantly boost overall output, but it can also create new wind-related stresses on the racks and systems. Then there's the issue of shading, which suddenly gets a lot more complex because you have both sides of the panel to consider, and that might impact the amount of spacing between rows. These issues are all worth taking into account, though, because the real energy gain in the field can be considerable, as much as 7%–9%.[25]

INTEGRATING THE GROWING
SOLAR RESOURCE INTO THE GRID

Shugar looks ahead and sees the solar industry continuing to grow at a rapid clip, propelled by new technological advances across the board, from the panels to the software. All of this will further drive down prices. And in some world regions today, solar is already the cheapest utility-scale resource. Recent contracts in the United States have seen prices below $20 per MWh—about half the cost of energy generated from a new gas-fired turbine. However, on its own, solar power is still at the mercy of the sun's trajectory, which means it's both a morning-to-evening resource, and one that diminishes significantly in certain regions during the shorter days of the year.

So how does one deal with that? Shugar responded that there are a number of arrows in that quiver. First, there is the stationary storage resource, largely in the form of batteries that are now being deployed on both sides of the grid—in the individual home and at the utility scale. Then, referring to his own Tesla, he said, "Vehicle-to-grid is a resource." You have to charge the vehicles at the right time, in the middle of the day, to absorb all of that extra power and flatten out the curve. Tesla's biggest car has a 100 kWh battery, "so my car could in theory carry over for days" serving the home with electricity, he said. "You could imagine a scenario where customers allow 20–30% of the capacity in the car to be used in the grid—that could provide a hell of a lot of value with no additional cost, as long as the charger is bidirectional."

Then there are other forms of modern demand management, such as turning off the devices at the customer site or altering their behavior in order to respond to the variability in solar energy. A water heater, for example, is conceptually just a very large thermal battery. And with Bluetooth it's getting easier to communicate with these devices, Shugar pointed out.

Ultimately, Shugar sees the industry's evolution at this point as being a steady trickle of technological gains, combined with ongoing professionalization: "We are just going to keep seeing more—I don't think it's like Eureka, and there's this breakthrough." He added, "I don't think about it like that—it's the cumulative effect of chipping away at a whole lot of things and we are kind of there now." He points to the classic technology adoption S curve, which has been getting steeper over time for technologies from telephones and cars to air conditioning, digital cameras, and the smart phone, with technology being adopted at an increasingly rapid pace. Shugar punctuated our conversation by referring back to the keynote he delivered at the virtual MidWest Solar Expo where our avatars had passed by each other. "One of the graphs I presented was how the pipeline looks. It showed that solar is more than everything else put together. So, the old shit is going to drop like flies over a hot fire."

MAKING A BIG BET ON BATTERIES

Other actors in the space are even more bullish on the opportunity for batteries. Leading publicly traded renewables developer NextEra, for example, indicated in April 2020 that it intended to invest $1 billion in batteries in 2021, and install enough storage capacity "to power the entire state of Rhode Island for four hours."[26] Shortly thereafter, the country's largest independent solar developer, 8minute (named for the amount of time it takes light from the sun to reach Earth) made an even more remarkable statement, commenting that it had over 18,000 MW of solar power and 24,000 MWh of battery energy storage under development in California, Texas, and the southwestern United States.[27] To put that number in perspective, 2018 saw 777 MWh of storage installed across the entire United States, followed by 1,113 MWh the following year.[28] Thus, 8minute is planning to install 20 times more batteries than the entire country did in 2019.

Dr. Tom Buttgenbach, 8minute's founder and CEO, tends toward seeing the future bigger picture, with a unique ability to frame the landscape in fascinating conceptual terms. I first spoke to him for an article I wrote in 2019, and in that single conversation, he shifted my perspective to a new way of viewing solar energy.[29] In his view, sunlight (and wind) are simply fuel that is concentrated in specific geographies of the world, much the same way hydrocarbons are.

Logically, the first projects locate in these new energy centers, and push out inferior and dirtier fossil-based technologies. In those hotspots in the southwest United States, solar will vanquish gas-fired peaking plants that run for a limited number of hours during the day when air conditioning drives energy consumption. Over time, though, as the costs of the solar technology continue to fall and efficiencies increase, the solar resource limited to those geographies begins—like an advancing army—to move north and east, vanquishing an increasing number of older, inferior, and dirtier technologies.

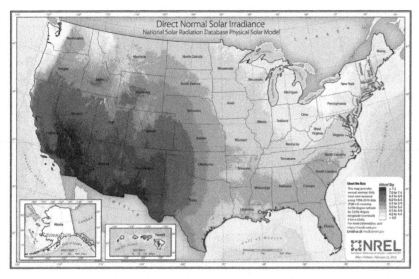

Irradiance, another word for where the free fuel is located. *National Renewable Energy Laboratory,* https://www.nrel.gov/gis/assets/images/solar-annual-dni-2018-01.jpg

When we spoke a year ago, Buttgenbach had discussed the potential of batteries as a future eventuality. Just a year later, the company was making a huge bet on the technology, and I wanted to know what had changed. So, I lined up a telephone conversation with him in late June to get his perspective on the rapid changes he saw taking place in the industry.

He commented that adding energy storage to the solar mix was the kind of critically important game changer that many people had yet to fully appreciate. Storage, he asserted, starts to put renewables on par with fossil fuel plants, which are considered to be the "reliable source" of power. But now, the combination of batteries and renewables are "substantially cheaper than fossil fuel," and "by adding the storage in there you are actually creating a much more robust and resilient grid than we have ever had in the past."[30] Buttgenbach likened the progression to digital cameras in a smartphone. "In the beginning, it was kind of cutesy and more of a joke. The picture quality was pretty dismal, but nowadays you have highly sophisticated cameras with slow-mo and nighttime and stuff so that you are hard pressed to buy a digital camera in the marketplace as good as what you have in your iPhone."

Likewise, he said, in the early days the critics of solar often commented that "yeah it's clean, but it's not dispatchable and it's not as available with clouds et cetera, et cetera." Today, he asserted, "The modern solar power plant like the one that we've designed as our new class of power plants—we have patents pending on the designs—it isn't just fully dispatchable. It can react within milliseconds and ramp to full power in a second." If there's an outage on the transmission line, for example, "this thing can jump in faster than your eye can see the lights flicker and make up for it. There is not a fossil plant in the world with these types of response times." He compares the new solar plants to fast starter gas plants that take 20 minutes to generate from turndown mode. "We guarantee 99-point-something percent reliability to the utility. You look at the fleet of gas plants in the Southwest and these guys are happy when they have more than 80% reliability. The difference between 99 point something that we guarantee and 80% is huge."

Buttgenbach added that in order to fully appreciate the implications, one really has to step back and see this from a completely different angle. "There's this complete paradigm shift in terms of how you think about reliability. With energy storage, you can firm up your local and distribution grid and create resiliency you haven't been able to do before." As a consequence, developers can look at optimizing the transmission system by strategically placing power plants and storage "in a way that both allows for the system to be optimized but also for upgrades, which is a big deal."[31]

With batteries, he noted, the rule of thumb today is to add four hours of storage capability for every megawatt of capacity added. Thus, a 1MW/4MWh battery could continuously yield 1 MW over a four-hour duration, for a total of 4 MWh. Or, it could provide a constant .5 MW over eight hours.

Buttgenbach indicated that adding the batteries can create a lot more value to the system in terms of flexibility and "dispatchability," since you can move the solar energy to any time of the day you desire. However, the addition of the battery also creates a far more complex planning environment, "because you can now design the solar system around the battery. The battery you can design to match the customer's load, and this is where the rule of thumb breaks down to 'what kind of problem am I trying to solve?'"

In the case of California, with its infamous "duck curve" (see chapter 11), Buttgenbach commented that it's blatantly obvious that batteries should be used in the evening to deal with the air-conditioning load that solar panels cannot directly address. But there's another little demand blip that occurs before the sun rises that also needs to be addressed in a market like California. "What happens when people wake up at 5:00 a.m. and the coffee machine and little electric heater go on at 5:00 a.m., 6:00 a.m.? At 8:00 a.m., the sun is up and the solar power plants begin producing solar energy. But for this one hour or two hours, the morning peak is a nightmare for a utility. Because it's, like, one hour long. You've got to fire up your gas fleet, which is quite costly with operations and maintenance . . . each start is very hard because of thermal stress from metal heating." There's not only the economics, but also the emissions to contend with. "As these turbines warm up, their NOx and SOx [nitrogen oxides and sulfur oxides] emissions are off the charts," Buttgenbach warned. But with a battery, that's not a difficult problem to address.

Each market environment across the county is different, and may well need a different solution. As a consequence, 8minute has different design criteria based on the location of the planned facilities and the problem to be solved. There's generally a trade-off between costs and outcomes, and while one can design for every problem, some solutions can become cost prohibitive. With batteries, the issues become more complex, but it has become abundantly clear that it's no longer just a case of the sun shines and the plant delivers power, whether the grid needs it or not. This new world of integrated solar and storage is "vastly more complicated than a simple solar plant. Now we need massive computer simulations and design optimization."

The other trend occurring in the industry is one in which developers are now contemplating overbuilding the plants to better integrate them into the grid. At first, this seems illogical *if* you are focused on utilizing every kilowatt-hour of solar energy produced. However, if the issue is how much solar can be cost-effectively integrated into an existing grid, the logic changes. With overbuilding, you have the ability to be flexible, ramping production up and down as requested (during the hours the sun is shining). Let's say you are committed to deliver 100 MW of solar. If you overbuild your plant, perhaps you can go up to 120 MW, or ramp back down to 100 MW. For certain hours of the day, that flexibility lets the solar plant offer the same cycling services offered by a flexible gas-fired facility.

Buttgenbach sees this scenario becoming increasingly commonplace, and employs the analogy of internet in the home. The typical customer may not use a gigabyte per second, "but it's so cheap that why not?" With cheap solar becoming increasingly cost-effective, you can now oversize the capacity. And if the PV plant is integrated with energy storage, it reduces the size and cost of the battery required to provide flexibility to the plant. "You now enter a whole new world," he observed, "where oversizing is a cost-effective way to solve the customer's problems and generate a lot more cheap energy. Now, what can we do with that additional energy to avoid curtailing it? It starts changing our whole way of thinking about this stuff, and the way the economy will evolve."

Buttgenbach sees a future in which some global economic activity may well be centered around regions of the world with vast solar (and wind) energy resources. He cited the example of data centers, which have traditionally been located in areas where they could access cheap electricity. They traditionally have been located in areas such as Virginia's coal country. Now, they go where cheap energy comes from renewables (in fact, entire countries such as Iceland and some of the Nordics, as well as Quebec, have been hanging out their shingles for data centers in recent years for just that reason). "It's a win-win," he said. "They look good in terms of signing solar contracts, but it's also saving them millions of dollars."

While Buttgenbach believes that electricity-hungry industries may increasingly migrate to the Sunbelt, lured by the promise of cheap power, he also thinks these energy centers will increasingly export cheap solar electricity to other parts of the country, if the country can get the transmission lines built. He uses the analogy of oranges: "How many oranges do we grow in Michigan? The answer is none. We move from where it makes sense to produce them and then bring those to the customer." He likened transmission to the U.S. system of freeways. "It would make perfect sense to build out transmission—just like the freeway system has connected markets together. That should be the next push for energy."

To Buttgenbach, the future of solar energy is limited only by our ability to conceptualize what it can do for society. "This isn't just replacing a gas plant. Once you have basically free energy, what can you do with that? That's where we are headed, and that's more than just replacing dirty fossil-fuel plants. It's fundamentally changing the way we work."

8

WHEN THE WIND BLOWS

If you've ever been to northwest Texas, you know that the wind almost never stops blowing, and much of the topography is pretty flat. In fact, several years ago while visiting that area, passing through the cotton fields up toward Lubbock, I commented on the sheer flatness of the landscape to my driver. "When you step out of my car," he suggested with a chuckle, "just look in the same direction for a while. If you wait a few minutes, you'll be able to see the back of your head."

It's not just the flat and windy terrain that makes Texas the fifth-largest wind economy on the planet, though. Rather, the key reason for that dominance is a unique public policy outcome that may strike some as somewhat unusual—at least for Texas. In 2013, under then Governor Rick Perry, the state completed a $6.8 billion ratepayer-funded investment in 18,500 MW of transmission lines stretching over 3,600 miles that linked up the wind-rich Competitive Renewable Enterprise Zones (CREZ) in western Texas and the Panhandle with the rest of the state.[1] In many instances, wind resources are located in areas where many people don't want to live, and transmission infrastructure is often lacking as a consequence.

Construction of those CREZ lines provided an opportunity for developers to come into the state and take advantage of the wind resource. And they came in droves. As of September 2020, the state boasted over 15,000 turbines totaling 30,900 MW of wind generating capacity, enough to power almost 8 million homes.[2] That number was up from 17,700 MW in 2015.[3] And as of September 2020, another 5,300 MW of Texas wind capacity was under construction.[4] (To put that into context, a typical nuclear plant might be sized at 1,000 MW of capacity.)

105

STRIKING WIND

Seven of those 15,000 turbines sit on John Davis's 2,500-acre Dry Creek Ranch in Texas, sitting roughly halfway between Austin and Midland/Odessa.

In early winter I had lined up a plan with Matt Welch, the head of Conservative Texans for Energy Innovation (CTEI), to fly to Austin and head out with Welch to the Davis's 140-year old Pecan Spring Ranch that sprawls across parts of Menard and McCullough Counties. There, the plan was to meet the Davis family, spend the night, and head out in the morning to tour the Dry Creek ranch—30 miles to the west—where the turbines from the Cactus Wind Farm spin.

Unfortunately, COVID-19 had other ideas for those plans, so eventually we set up a teleconference call instead, where Welch and I spoke with three generations of the Davis family. Speaking to us from their couch, John Davis was surrounded by his father Jim ("Pappy"), the 90-year old patriarch of the family, and youngest son Gaston ("Stoney") Davis.

Davis is not your bleeding-heart, clean-energy obsessed liberal—far from it. For many years he owned an industrial roofing company, and served 16 years as a Republican member in the Texas Legislature, chairing the House Economic and Small Business Development Committee for two terms.

Like many smaller ranchers these days, the Davis family's connection to the land runs deep, but that alone won't pay the bills. One has to be flexible and tap multiple revenue streams in order to make a go of it. The family labors to keep both ranches intact, a task requiring agility and a commitment to do what needs to be done. John, his wife Jayne, and three grown sons all do what they can to make ends meet. That includes "Jayne's Specialties"—jams and preserves tailored from ingredients grown at the ranch, and specialty meats that are both raised and processed on the ranch.

At the Pecan Spring homestead, a small herd of both Wagyu and Aberdeen cattle—prized for their richly marbled meat—share the ranch with Angora goats that yield mohair. Up at Dry Creek, close to 300 head of Dorper sheep, some Spanish and Savannah goats, and a handful of cows roam the property. The market has generally been kind. Customers like their products and the story that goes with buying food from a ranch over a century old. However, COVID-19 has been less forgiving, and the Davis clan has taken to selling more locally since the Houston farmers' market was interrupted.

Meanwhile, Davis said, there are always animals to feed, equipment and buildings to keep up, and the relentless bills and taxes. "The cold hard fact is you gotta generate money. Whatever it takes, and however you do it. You piece a living together from whatever works."

The Davis family is fortunate and possesses one advantage some other ranchers do not: their land in Concho County is rich in fuel. However, it's not the hydrocarbons that have led to more boom-and-bust cycles in a place like Odessa than anybody can count. Instead, the fuel on their ranch is wind, and it hardly ever stops blowing.

Five years ago, a developer from RES approached Davis about putting some turbines on his land and, as he puts it, "We struck wind." RES had been prospecting in the Concho County area, at the same time landsmen from other companies were scouring the best locations in Texas, Oklahoma, and much of the entire wind-rich spine of the country that runs from Texas all the way to the Dakotas. They and others like them were erecting meteorological towers, assessing wind resources, evaluating the best locations, sizing up critical grid interconnections, researching property deeds, and approaching landowners with long-term agreements to erect turbines on their land.

A large number of landowners were understandably reluctant to allow the siting of turbines on their properties, and there has been more than one contentious dispute between neighbors for and against the turbines. The towers of the 3.45 MW Vestas turbines at Dry Creek are 80 meters high (262 feet), and the blade soars another 64 meters (210 feet).[5] The machines emit an audible hum, and their warning lights blink all night. Then there are also the access roads and structures to house equipment that dot the landscape.

Pappy was among the doubters at first, and not too keen on the whole idea. However, the rest of the family appreciated the financial cushion the annual royalty payments from seven turbines would offer, helping them to keep the land intact for the foreseeable future. As we conversed on our respective computer screens, John Davis looked over at Pappy and gently reminded him, "Dad, you said we could do it. We kind of wanted your blessing for it before we moved forward." He looked at us on the Zoom screen and said, "Dad gave the blessing, said OK. The same thing happened to other landowners. They had other generations. . . . It's a way we can keep our heritage alive."

The family negotiated a contract, covering issues such as royalties, reimbursement for any damage to the land, and decommissioning and

The Davis men on horseback at their ranch in Concho County, Texas. *Photo courtesy of the Davis family*

removal of the turbines upon contract termination decades later. Eventually, the 43-turbine, 148-megawatt Cactus Flats Wind Facility project was commissioned in 2018, and royalty checks began to flow. (The wind farm generates electricity that serves a General Motors plant in Arlington as well as General Mills facilities.)

The Davises are not the only ones in Texas reaping the financial benefits of wind. The American Wind Energy Association estimates that over $50 billion has been invested in wind projects in the state, creating somewhere around 25,000 jobs. Annual land lease payments to landowners in the state exceed $192 million, while state and local tax payments total roughly $285 million.[6] In other words, wind is big business in Texas.

This holds true for that entire windy area of the United States. Texas gets about 17.5% of its total electric energy mix from wind, but some states are far higher in terms of their relative contributions. Iowa topped the charts at 42% in 2019,[7] with Kansas at 41%,[8] while Oklahoma trailed slightly behind at 35%.[9] For the entire country, that total was 7.3%.[10]

BIGGER, BETTER,
AND MORE POWERFUL TECHNOLOGIES

Some areas of the country are blessed with the kind of incessant wind that can drive a person crazy while lining one's pockets with royalty checks, while other areas, like much of the southeastern United States, have more limited wind resources and few turbines to date. That may change in the future as turbine technology improves and the machines get larger.

That's because turbines are getting larger and taller, which allows two things to happen: they can access the stronger wind currents that exist further off the ground, and they are also more efficient at harvesting that wind. An increase in tower size[11] potentially opens up much of the rest of the country to cost-effective deployment of wind turbines.[12]

Wind Capacity by State

Where the wind resource is being accessed. *American Wind Energy Association,* https://www.awea.org/wind-101/basics-of-wind-energy/wind-facts-at-a-glance

Wind turbines are in fact continuing a steady growth trend. In 2015, the average turbine installed in the United States was 2.0 MW.[13] By 2019, that number had climbed to 2.55 MW,[14] with fully 25% of the turbines exceeding 3 MW. The tallest onshore project to date currently stands at just under 200 meters (654 feet) from the base of the tower to the tip of the blade[15] (soaring more than twice as high as the Statue of Liberty's paltry 305 feet).

JUST THE BEGINNING

Yet, as far as size goes, the industry may just be getting started. Aided in part by efforts to develop the huge offshore turbines that are now up to 15 MW in capacity, some onshore machines currently exceed 6 MW. The three major western manufacturers—GE, Siemens Gamesa, and Vestas—now all offer onshore turbines well over 5 MW (to give one a sense of the size of the machine, the "swept space" covered by the blades is just under five acres). In June 2020 Vestas made public an inaugural U.S. order for 45 of its 5.6 MW machines for an installation in Knox County, Texas.[16] In the same month, Siemens Gamesa announced its first U.S. order, with 66 of its 5.0 MW machines dedicated to an unidentified 330 MW project in Texas.[17] To put the size issue in perspective, the total rotor diameter of the Vestas machine is 162 meters. That would be more than two-and-a-half Boeing 747s parked wing tip to wing tip.

The race is on for these bigger machines because they deliver more electricity at a lower cost. To understand why size matters so much, it helps to understand the basics of wind energy: Unlike solar energy, where the light from the sun harvested by panels is fairly predictable for any location, wind is much more dynamic.[18]

HOW WIND TURBINES WORK

The turbine and its three blades function like an airplane propeller or helicopter rotor. Wind moves across the blade, creating a difference in air pressure (from the top surface of the blade to the underside) that produces lift and causes the rotor to spin. The blades of a wind turbine don't get pushed by the wind. Instead, they fly in the wind much like an aircraft. The rotor is connected to a generator with a series of gears that increase the rotation of the rotor from 12 to 18 revolutions per minute to approximately 1,800 RPM in the generator. It produces AC electricity at 60 Hz that can be synchronized to the grid.[19]

The actual output of that turbine is affected by three variables: air density (because that creates more or less "lift" on the blade); the swept space (the wider the rotor area covered by the blades, the more energy one captures); and the speed of the wind. Of these elements, the one that matters the most is the wind speed, since the output of a turbine is a function of the cube of that velocity. Thus, a steady 20 miles-per-hour wind will yield eight times the energy of a 10 miles-per-hour flow.

Taller turbines allow one to access stronger and more consistent wind currents that are less affected by the earth's topography. For their part, longer blades enable the turbine to harvest more energy. The machines themselves need to deal with a harsh environment over decades, with enormous variability in winds and significant stresses. So the turbines are equipped with controls that can change the angle of the blades in response to wind gusts—the blade can literally rotate away from the wind and change its angle of attack. At the same time, the turbine is always facing into the wind, so it needs to be able to rotate (yaw) on its vertical axis.

The average turbine starts to generate power when the wind speed is between 6 and 9 miles per hour (the "cut-in speed"). Then as the velocity of the wind increases, the turbine reaches full rated capacity at approximately 20 miles per hour, after which the turbine feathers the blades and begins to dump wind. The full-rated capacity is the maximum generator output that defines the MW rating for each turbine model. When the wind reaches speeds that could damage the turbine, usually at around 40 miles per hour, the turbine will shut itself down and wait for the storm to pass.[20] In the most extreme cases, in gusty conditions, the blades can even be locked down to protect them. For large machines with longer blades, tip speeds at the end of those blades may exceed 200 miles per hour.[21]

LOOKING FOR GAINS ACROSS THE ENTIRE SYSTEM

Turbine manufacturers are constantly seeking improvements, from the blades, towers, turbines, and software to the transportation of the assets themselves. The CEO of the American Wind Energy Association, Thomas Kiernan, told me that "we are still a young industry and seeing productivity gains in production and cost. Longer blades, taller towers, and digitalization are the three biggest dimensions to further cost reductions and productivity improvements."

With blades, the critical challenges are making them longer, lighter, and more durable. Among the solutions being developed are increasing the use of carbon fiber and high-tech glass fibers. In addition, aerodynamic designs continue to improve, resulting in lighter and more slender blades.[22] One challenge, however, is that longer blades—especially those beyond about 65 meters (213 feet) in length—are harder to move from factory to site, due to existing infrastructural constraints such as bridges and tunnels. One way around this may be employing creative approaches such as GE's strategy of developing two-piece blades.[23] Another may be fabricating blades on-site—so-called gypsy manufacturing. And yet a third way may be employing dirigibles to move bulky blades over the landscape.

The towers face their own transportation-related constraints. They're also difficult to move across the countryside, so some alternatives are being proposed, including spiral welded towers that may eventually involve production of super-tall towers on-site.[24] Hybrid concrete and steel towers are already being employed in some instances.[25] And in June 2020, GE announced a partnership with concrete giant Lafarge Holcim and 3D printer manufacturer COBOD to print wind towers as tall as 200 meters (656 feet).[26]

Other system improvements are being made as well, with a significant focus on the application of digital technologies. Here, the key is to optimize the design of the turbines to both the landscape and the constant operation of the entire wind farm. One critical element that digital tools allow operators to improve on is the issue of wake. Every turbine harvests energy from the wind, but the downstream current is much like the wind that spills off the sail of a boat—turbulent and disorganized.

In a typical wind farm the turbines are spaced far enough apart (often well over a third of a mile between turbines in a row and more than a half-mile between rows) so that they are not constantly interfering with each other's operation when the wind comes from the prevailing direction. However, there are times when the wind comes from certain directions, and the turbines are in close enough proximity, that output resulting from wakes can be decreased by as much as 40%. That issue can be largely addressed by a technique known as "wake steering," in which the turbines are yawed slightly away from the prevailing wind direction so that wakes are deflected away from downstream turbines. Tests applying industrial algorithms to optimize the wake steering demonstrated increased output for certain wind directions, up to 7%–13% for moderate winds.

Just as with solar and battery technologies, it's patently obvious that wind technology still has the potential to continue climbing up the productivity curve, resulting in lower-cost electricity.[27]

NEXT ON THE HORIZON

At the end of the third quarter of 2020, there were almost 60,000 turbines in 41 states and two U.S. territories, providing 110,800 MW of capacity,[28] equal to 7% of U.S. generation.[29] Tom Kiernan, the American Wind Energy Association CEO, thinks the industry will build a whole lot more.

Kiernan is bullish on the future because, as he points out, "Wind is the cheapest source of new electricity in most parts of the country, so this transition and transformation of the energy economy is inexorable. It is happening, it's going to happen faster, and it's driven by low cost and clean energy that people want . . . the transition is in its earliest stages and will be accelerating."

Kiernan identified some challenges that must be addressed if the U.S. wind industry is to meet its full potential. "Transmission is first," he noted. More high-power lines will allow more wind energy to be generated in the windiest regions and moved to markets that need it. "Policy is second, and siting is also a growing challenge for the industry." Kiernan called out opponents who "are providing misinformation and coming into communities from other states or from out of the region. It's important for the industry to create some clear best practices and standards backed up by science and analysis so that communities can be comfortable and best practices are fully analytically justified."

Kiernan argued for continued stable policy support at the federal level, citing the "near-permanent support for fossil fuels far beyond what renewables have received." That can be achieved either by putting a carbon tax on fossil fuels to account for their negative attributes, or subsidizing renewables, so long as there's a level playing field.

Kiernan also pointed to the fact that companies such as giant renewables developer NextEra are already combining wind and solar and battery projects—such as the 250 MW wind, 250 MW solar, and 200 MW (four-hour duration) battery Skeleton Creek project in Oklahoma, and a similar hybrid project in Oregon. "If one adds batteries and software to the right mix of wind and solar, you have a fairly reliable asset."

By adding those technologies and improving forecasting and scheduling tools that allow firms to better deal with its output variability, wind can constitute an increasingly large percentage of the energy generated in this country. Kiernan observed that "20% by 2030 is our goal," and stated, "We envision well over 50% by 2030. Where and when we go from there, we will see. The key point is the next 10 years, it's pedal to the metal. And the grid can absolutely handle that reliably."

Just as with any other industry, education and lobbying are important, Kiernan explained, and his group is constantly providing industry information "so folks have something to push back against those saying wind energy causes cancer. We need to have the information documented and distributed." And, he noted, it's critically important to get the word out about the economic benefits—not just in Texas, but across the country.

Last year, his association did an analysis and found that the wind industry supplies roughly a billion dollars a year to rural America, including about $750 million in taxes and other county payments and about $250 million in land lease payments. Unlike many issues these days, wind can be bipartisan. In fact, he pointed out that "80% of the wind farms developed since 2016 are in states the president [Donald Trump] carried. There are a lot of Republicans that see the benefits of wind energy—this is very much a bipartisan supported industry."

During our discussion, Kiernan also speculated on the future potential for using wind energy to create hydrogen. Not three months later, a report from Morgan Stanley indicated that surplus wind may be used to create cost-competitive hydrogen in the U.S. Midwest and Texas within the next two years.[30] He is optimistic about that as a possibility, but was quick to point out in our conversation: "A key observation I would make is clearly the renewable industry broadly needs to keep innovating and driving new technology, but it's not as though we need a new technological

Four generations on the land. *Photo courtesy of Jim Davis*

breakthrough or step change for this to work for the country to make this economical. These are all technologies we know that can work and are making a difference now, are economically competitive now, and will only get more so in the coming years. We don't need some huge breakthrough."

In the meantime, the Davis family and others like them collect the royalty payments that can help keep their properties and lifestyles intact so they can be passed down to the next generation.

Pappy Davis, the family patriarch who grew up on the ranch and spent decades of his career as a landsman signing contracts in the oil industry, is hopeful that these new renewable energy resources will help keep the land in the family for generations to come. Davis summed it up in our virtual face-to-face conversation across 2,000 miles of distance. "When the wind blows," he said, "it's money."

9

STEEL IN THE WATER

In August 1924 the whaling bark *Wanderer*, the last of the once-vast fleet of square-rigged whaling vessels out of New Bedford, Massachusetts, lifted anchor and left the safety of the harbor. It was an ill-timed departure. Less than a day later, she was driven by a murderous gale onto the treacherous Middle Ground shoals off Cuttyhunk Island.[1] While the crew was saved, the *Wanderer* was a total loss—the final remnant of America's first large-scale foray into an international energy industry. Centuries ago, it was the wind in their sails that drove the fleet of whalers around the world on lengthy voyages in the quest for oil. Today, the wind that propelled that whaling fleet is the new source of energy we seek.

The old New Bedford Harbor was populated with a forest of ship's masts and spars. If all goes as planned, that same port soon will be bustling with cranes lifting heavy nacelles and massive wind blades—three times longer than an old-time whaling ship—onto specially designed vessels for another sea voyage. The U.S. offshore wind industry is on the precipice of becoming enormous, worth tens of billions of dollars. In Europe and China, it has already existed at that scale for years.

I wanted to learn more about it, understand more about some of the equipment, and get a sense of how the industry was evolving. So, on a frigid February morning in 2019, I piloted my Hyundai Ioniq electric vehicle (with the heat off—I needed all 130 nominal miles of range I could get that day) into Charlestown, Massachusetts, an urban cluster just outside of Boston, to visit the Wind Technology Testing Center (WTTC). I left my license at the guard shack in the shadow of the hulking green Tobin Bridge spanning the Mystic River and navigated my way past rows of imported cars (fruitlessly hoping to spy a few electric vehicles to keep my

Ioniq company), made a few wrong turns, and eventually found myself at
the giant gray steel shed.

TESTING THE EQUIPMENT

The only indication of the activity inside was the presence of enormous
white letters spelling out the name on the side of the building and two long
and sleek blades—well over two-thirds of a football field long—resting in
front. Covered under tarps for competitive security reasons, they sat waiting to be put through their paces.

Across the channel to the north, a large black liquid natural gas (LNG)
vessel was berthed, accompanied by a state police escort boat, its blue lights
blinking as the ship unloaded its cargo of natural gas to the Distrigas terminal in Everett. Behind it, the twin emission stacks of the soon-to-be-closed
Mystic Power Plant loomed, while to the west, the blades of a 1.5 MW
land-based turbine whirled silently. That scene represented a good cross-
section of today's energy world. I got out of my car and entered the building
to get a better look at the energy landscape—and seascape—of tomorrow.

WTTC Executive Director Rahul Yarala warmly welcomed me back
into the upstairs conference room where we grabbed a cup of coffee and
caught up. I had last visited him to profile his facility for an article I wrote
in 2013.[2] A lot has changed since then, and Yarala has seen the industry's
evolution manifested in the constant parade of bigger and better blades
marching through his facility, so I was anxious to hear his perspective about
where the industry is headed. Yarala has been involved with the WTTC
since he was hired to work with the initial design team, helping to establish
the required parameters.

Some Surgery Required

A window in the conference room where we spoke looked out into
the facility, cluttered with toolboxes, a few Genie lifts, huge steel rings
waiting to be affixed to blades, dollies, and cut plywood forms in the shape
of the cross-section of various-sized blades. With the exception of the two
yellow 50-ton KONECRANES that perched on reinforced rails, nearly
the entire facility was steel or gray. "Cost-efficient" was the phrase that
came to mind.

Dominating the view was an enormous blade that stretched across
most of the entire 90-meter span of the facility. This was just part of the 12

A small blade being tested. *Photo courtesy of the author*

MW, 107-meter GE Haliade-X test blade that had been shipped over from its manufacturing facility in Cherbourg, France. Offloaded from a vessel by the ship's own cranes, the blade had been transported along the quay on two separate electronically controlled and self-propelled mobile transport units, crewed by a specialized team imported from New Jersey.

Then, because the shed is 17 meters shorter than the blade, the team had sawed through the blade—a combination of fiberglass, specialized composite materials, and balsa wood frame—leaving the 70-meter base section to be tested (the process of cutting did not affect the type of fatigue testing to be done). They then hoisted it with the WTTC's cranes and bolted it to a massive steel plate on the specially engineered wall, which can accommodate a blade root as wide as 5.5 meters.

The day I visited, the WTTC team was in the process of setting up the blade for the testing process, and they had already initiated the process

Adding the sensors. *Photo courtesy of the author*

of attaching the 200-plus sensors and strain gauges with their trailing wires, meant to pick up indications of fatigue or any other abnormalities during a testing process that would simulate two decades of harsh marine weather and loading from the extreme wind and waves.

Over the course of the standard testing procedure, a blade is subjected to as many as several million cycles of flexing—side to side, as well as up and down. The process can take a few months for shorter blades, and up to 8–10 months for the longer ones, a highly accelerated life test representing 20 or more years in the field and set to the international standard IEC61400-23.[3] Even the planning process to set up the blade properly for the test can take many weeks. The price tag for this fatigue testing ranges from $500,000 to $1 million, depending on variables such as the size of the blade and the procedures specified.

Yarala and I donned hard hats and walked out to the floor to stand beneath the GE Haliade blade. Looking up, I was impressed by its sheer size and unusual color. With its size and gray-green hue (in the field, these are coated with a special white paint), it reminded me of the skin of a whale that had been through a rough patch or two during its life. But looking out toward the tip (or where the tip would have been, had it not been sawed off), it evolved into a more refined and tapered curved form.

These latest blades and the turbines are technological miracles, and the fruit of a long and patient evolution over decades. It's difficult to grasp the sheer enormity of these machines. To start, let's talk size. The 40-ton, 107-meter (350-feet) blades,[4] sitting atop 260-meter (850-feet) towers,[5] harvest the kinetic energy of the wind. The "swept space"[6] capturing that energy (i.e., the space between the outstretched arms, from blade tip to blade tip) covers 38,000 square meters (410,000 square feet or 9.4 acres), roughly equal to the area of seven football fields.

When the wind is cranking (the optimal speed is around 23 miles per hour[7]), the 12 MW Haliade rotor can generate about 450,000 foot-pounds of thrust at full power, the equivalent of the power necessary to launch a four-engine Being 747 off the ground.[8] In fact, a *single rotation of the blades generates enough energy to power the average U.S. household for an entire day* and one turbine can generate enough energy annually to supply over 5,000 U.S. homes with all of the electricity they need.

The Groundwork Was Laid Over a Decade Ago

While these huge wind turbines are the result of decades of work and planning, so too is the mere existence of a budding U.S. East Coast offshore wind industry. The behind-the-scenes effort to create the infrastructure necessary to support this multibillion-dollar industry that will provide thousands of jobs up and down the East Coast has been well over a decade in the making. It involved a lot of faith, and willingness to invest both money and political capital in a future that was—at the time—anything but certain.

As we talked over coffee in the conference room, Yarala explained that he and colleague Eric Hines, Tufts University Professor of Practice and director of the Offshore Wind Energy Graduate Program, began to investigate the need for such a testing facility way back in 2008. Their hope was that the WTTC might be funded by an industry that was almost solely focused on onshore wind at the time. However, before they could line up private financial support, the financial markets collapsed in the Crash of 2008. Fortunately, Obama administration stimulus money came through, and a combination of state and federal money totaling close to $40 million ultimately helped get the facility built.[9]

Even as they were working to line up the funding, the small team of innovators was asking the question about what, exactly, needed to be built, and what could be achieved with a budget that wasn't exactly flush with cash.

Yarala told me that the first step was to go out and survey various company representatives and find out what they needed and what the industry would support.

In a separate conversation, Hines told me that there was a great deal of initial skepticism both from those in the public sector and, indeed, from some in the industry. "When we first put together the WTTC concept, we got comments saying, 'Nobody's going to use that thing,' and now it's fully loaded," Hines said, with all three test stands in operation most of the time.

Yet while skeptics abounded, Yarala said, there were others in the wind industry that very much wanted such a facility, partly because they were gun-shy of the optics related to potential failure in the field. It's not hard to go online and search for spectacular videos of the early wind turbine failures, the turbines on fire spinning crazily into oblivion, collapsed towers, and broken blades. Those incidents still occur on occasion, but far less frequently. However, unlike the coal- and gas-fired generating plants, Yarala noted, where failure is generally not visible to the public, if a wind turbine fails, it is not only visible, it is *very* visible. And—especially in the industry's early days—when the industry had not yet proved itself and there was a good deal of skepticism, a failure of one manufacturer could be extrapolated to the rest of the players, giving the entire industry a black eye.

Today, Yarala noted, the reliability is high and failure less likely, but a continuous improvement in the economics of wind energy requires ever bigger machines that can sweep more space in the sky and reliably generate more energy. Thus, he says, "you have to have the infrastructure for new ideas to be developed and tested."

The One Thing They Didn't Anticipate

A lot has changed in the roughly 2,500 days since my 2013 visit, he said, with nearly the entire focus of the center having shifted to testing offshore blades. That evolution was relatively rapid and unexpected. When the WTTC opened, it began testing blades in the 45- to 50-meter range, largely for onshore projects. However, it soon started to test "a bunch of 62 to 65 meters, and now we are seeing around 70 meters for onshore." And then of course there is GE's 107-meter behemoth, which will be followed by others even larger in the years to come.[10]

This is—for now—the new frontier, with offshore turbines roughly twice the size of land-based machines. The challenge, as Yarala puts it, is that "on the drawing board, in the computer, you can make it bigger and cost-effective, but you need to ensure the manufacturing is there, the reliability is there. These are massive structures that have to be built reliably."

Furthermore, the blades are simply too large to be manufactured solely by machines. "With a wind turbine blade," Yarala commented, "you cannot build a tool to automate manufacture of a 40+ ton, 107-meter-long blade. That would be like saying you could build a football stadium somewhere and then drop-ship it. You can automate construction of a modular one-room home, but not all of Gillette Stadium."

These high-tech but still largely artisanal blades, with their own unique construction dynamics, must be thoroughly tested. And Yarala commented that the evolution to increasingly larger blades is dynamic and not likely to stop anytime soon. "We've been approached by and are in talks with Siemens Gamesa and Vestas to test longer blades, . . . The DOE [U.S. Department of Energy] just released a 15 MW plan, and in Europe they just released plans for 15–20 MW machines. So now, we're are talking 120-meter blades, that will be even longer and heavier."

All that is to say that in 2008, Yarala and Hines surveyed the industry, extrapolated growth well beyond anyone's expectations, and eventually wound up with the funding for a state-of-the-art, 90-meter-long sophisticated testing facility that was almost twice the length of the blades they were then testing. And now, just nine years in, the facility is looking just a bit too small.[11] That's the pace of change being seen in the industry.

With the GE blade, Yarala said, "We cut the tip path, because the last part of the tip is not that critical from a structural testing perspective. We can test the first 70 meters and confirm the structural integrity, or we can test in two pieces." He commented that research is underway to confirm whether a future protocol that involves testing multiple segments—rather than the entire lengthy blade—will be adequate. "As an engineer, I say 'yes,' but as a pragmatist I say, 'wouldn't it be better if you could test them in one piece?'"

Testing to Failure

The actual testing procedures were not taking place during my visit. Instead, the team was focused on setting up the GE blade over the course of three to four weeks, attaching the sensors and strain gauges along the length of the blade. During testing, these will transmit data back to computers equipped with specialized software that translates the raw information into stress levels and other important indicators, constantly searching for anomalies.

None of that is a substitute for the human eye, though. Yarala noted that every second day during the testing sequence, his engineers will climb into the belly of the beast, crawling inside the hollow blade and searching for anomalies that should not be there, such as changes in the blade material color. "A human expert can see whitening of the surface and see potential failure quicker than a gauge." That will all eventually improve with artificial intelligence, he commented, and cited ongoing work with University of Massachusetts–Lowell and other universities to develop these capabilities.

The entire building itself was constructed to withstand the constant stress of testing. The facility may look like a prosaic oversized warehouse, but it is in fact a marvel of engineering in its own right. The point where the blade root is bolted to the wall is the main focal point for stress, so Hines, Yarala, and the team designed it to include tons of concrete and reinforced steel. Among the most critical design issues was the ability of the facility to withstand both long-term fatigue (100 million cycles) and short-term stresses. As a consequence, the hybrid foundation includes 15 concrete shafts 4 feet in diameter and 170 feet long, so the applied load is transferred to the solid rock below.[12]

That stress is greatest during the process that Yarala referred to as "test to failure." The customers determine in which direction they want to stress the blade one final time, and they show up for this test without fail. "We've done 15 or 20. It's a special event," Yarala commented dryly. "As director, I'm always worried about what it's going to damage. But as an engineer, you do the calculations and say 'we are going to be OK.'"

During this procedure, winches attached to the floor begin cranking the blade down, bending it further and further, while the team observes the "event" from behind shatterproof glass. Depending on the blade, and the testing protocol, the blades fail in different ways. Sometimes, they simply break, and the failed section dangles, which may sound somewhat anticlimactic, says Yarala, "but at the end of the day, 15 or more tons of mass is coming down, whatever the cause of failure." And then, there are those more spectacular moments when the potential energy rapidly converts to kinetic energy, with the accompanying noise. "In a microsecond, boom. . . . One of the times, the guard shack (which is easily two-tenths of a mile away) called us" to check if everything was OK.

THE BIGGEST TRANSITION
IN THE HISTORY OF THE WORLD

Today, estimated capital expenditures just for the state offshore wind commitments announced to date are in the $70 billion range, and more will undoubtedly follow.[13] Development of such a massive industry doesn't just happen instantaneously. It takes foresight, planning, and resources.[14]

In fact, the story of today's East Coast offshore wind projects goes back over 15 years to 2005, when the federal Energy Policy Act tasked the U.S. Bureau of Ocean Energy Management (BOEM) with the responsibility for renewable energy development on the outer continental shelf. Just four years later, the Massachusetts government began collaborating with the BOEM to identify and establish specific leasing areas, as close as 14 miles off Martha's Vineyard and extending well offshore. The development of these leases included evaluation of shipping routes, ocean habitat, and fishing grounds. In the beginning, there was not much industry interest. Offshore wind was being tested in Europe, but the prices delivered to that marketplace were extremely high, and well out of the range that anyone was willing to subsidize.

Nonetheless, state officials began planning for a future they hoped would happen, undertaking studies from everything related to the necessary workforce development, to the critical ports and infrastructure, to the transmission capabilities required to integrate a massive infusion of offshore energy into the grid.

That meant that over the course of the years, there were many highs and lows. The failure of the very first proposal, Cape Wind, a $2.6 billion, 468 MW, 130-turbine project, was one such low. It would have been built close to shore and visible to the playground of the rich on Nantucket and Martha's Vineyard, and as such it ran into the buzz saw of well-funded public opposition.[15] And yet according to Bill White, who played a key role in developing the Massachusetts offshore wind market during the Massachusetts gubernatorial administrations of Duval Patrick and Charlie Baker, despite its eventual failure, Cape Wind clearly was the catalyst that served to educate policymakers with some painful lessons learned while providing a roadmap for the required infrastructure needed to not only realize projects, but reap the clean energy and economic benefits associated with the new industry.

MORE THAN AN OVERSIZED PARKING LOT

Hines pointed to the New Bedford Marine Terminal—initially developed to support Cape Wind as its first project—as another critical infrastructure investment that took a good deal of faith and planning. Today, he said, it looks "like a huge parking lot," but in fact will soon serve as a strategic gateway for the development of much of the northeastern U.S. offshore wind industry. It has enough space to support storage, assembly, and loading areas, and the movement of thousands of tons of equipment. It's well connected to the local highway system, and it is now capable of handling the extremely heavy loads soon expected to come its way.

Unlike Europe, which benefits from the existence of ports such as Bremerhaven, Rotterdam, and Liverpool[16] that have served the massive offshore oil industry for many years, the U.S. East Coast does not possess a single deep-water port with the infrastructure to support the offshore wind industry. All of that infrastructure was either built in recent years, or must be built soon to support the industry as it forges ahead from Maine down to North Carolina.

Recognizing that challenge, in 2010 Massachusetts commissioned an analysis to identify local port facilities with the ability to support the future offshore wind industry, as well as explore the feasibility, economic development potential, and economic impacts related to development of potential ports.[17] The report evaluated the weight and size of the potential equipment to be deployed, as well as related navigational issues, ranging from flight activities at Logan Airport in Boston to the height of bridges that might limit transport of the equipment—most notably the bulky blades—and required channel depths.

New Bedford was eventually anointed the winner, and then the real work had to be done, building the 29-acre facility, extending the bulkhead to 1,200 feet to take deep draft vessels up to 800 feet as well as 400 feet of space for barges, and reinforcing the ground to accommodate loads of 4,100 pounds per square foot with crane loads up to 20,485 pounds per square foot.[18] The project was not cheap, with a price tag eventually coming in at $113 million.[19] Timing is everything. Completed on time in 2015 to support the ill-fated Cape Wind project, the terminal has not yet met its potential, with the exception of visits by several vessels, including a few to offload wind equipment for a project in Massachusetts. However, in recent years the facility has worked to break even by hosting multiple offshore wind industry vessels that are working to survey the sea bottom floor. Developer Vineyard Wind has a $6 million per year lease to use part of the space for 18 months commencing in December 2020, but its project has been delayed due to an unanticipated slowdown in the supplemental federal offshore environmental review process that has slowed the project by over a year.

Yet White, now president and CEO of EnBW North America (the subsidiary of the German utility and offshore wind developer EnBW AG), noted that the industry is about to take off, and the forces are massing to make it happen. As of early 2020, East Coast states had committed to over 29,000 MW of offshore wind procurements.[20] When the industry finally does take off, with thousands of workers converging on bustling ports and huge equipment mobilized and moved, the realization will dawn on the general public that we have an entirely new and substantial industry just off our shores. White is hopeful that the scope and scale of this industry will particularly inspire the next generation: "When the fifth-grade science classes gather on the New Bedford Hurricane Barrier or on the point of Staten Island and witness these massive turbines being mobilized on vessels, it will capture the imagination of the next generation of scientists, engineers, and tradespeople. That generation, who feel the urgency of climate change like no other, will see the clean energy future."

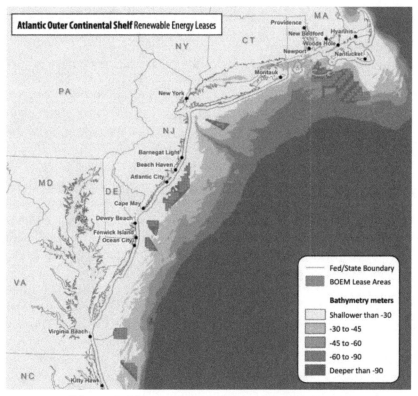

East Coast offshore wind leases. *Bureau of Ocean Energy Management,* https://www.boem.gov/sites/default/files/renewable-energy-program/Mapping-and-Data/Renewable_Energy_Leases_Map_Book_March_2019.pdf

Eric Hines noted that a review of the lease prices shows just how quickly market perception of the opportunity has changed. The BOEM has already executed 13 competitive leases since the inception of the program, from Massachusetts down to North Carolina.[21] Consider the fact that in the January 2015 BOEM auction, only two parcels attracted bids of approximately $260,000 (from RES, which subsequently sold to Ørsted) and $168,000 (from Offshore MW, which subsequently sold to Copenhagen Infrastructure Projects).[22] Four years later, leases for parcels about 30% larger in similar areas went to three separate parties—Equinor Wind, Vineyard Wind, and Mayflower Wind—for $135 million in each case.[23] Those same record-setting lease areas had attracted no bids just a few years earlier.[24]

IF IT'S OFFSHORE AND INVOLVES HEAVY INFRASTRUCTURE, IT'S OUR BUSINESS

What kind of player can make that kind of speculative bid and even play in such a capital-intensive game? Unsurprisingly, in many cases, these entities are the same European companies that have been braving the elements and turning profits in offshore oil and gas development for decades. Whether it's a hydrocarbon underwater and underground, or free fuel from the wind, these large companies know how to manage the capital, helm the ships, and move the equipment to get steel in the water.

In order to understand how this industry will likely develop, it's worth taking a look at a few of the key players. Ørsted, formerly known as Danish Oil and Gas, is a global leader in offshore wind with a 25% market share and the winner of 2,900 MW of U.S. offshore project awards to date.[25] Equinor is a Norwegian state-owned company that used to be known as Statoil until mid-2018.[26] It was recently awarded the right to develop the $3 billion, 816 MW Empire Wind project to deliver energy to New York.[27]

Shell and EDP Renewables (a Spanish company specializing in large renewable projects) recently won the Massachusetts bid, beating out two competitors to proceed with Mayflower Wind in a joint venture. This 804 MW project turned some heads with its record-setting price of $58 per megawatt-hour over the contract duration. That was just under 25% of the $244 contract price for the Block Island, Rhode Island project, the first U.S. offshore wind project, commissioned in 2018.

Vineyard Wind is one of the few projects not being developed by a hydrocarbon giant. Its joint-venture owners are Iberdrola (a large international utility holding company based in Spain) and Copenhagen Infrastructure Partners (a fund management company run by former energy executives, overseeing $7.5 billion of investments in offshore wind, other renewable technologies, and offshore electric transmission).

BP is late to the game, only recently designating offshore wind at the top of a list of key technologies and sectors "set for more mainstream success." To date, it has limited exposure to an older fleet of U.S.-based turbines, but may finally be about to dip a toe in the offshore waters.[28] French oil major Total has also recently entered the game, though it has yet to win a bid. Conspicuously absent in the mix to date are the U.S. oil majors and drilling companies. Surprisingly, with a bonanza exceeding $70 billion, none of them have shown up yet to claim a stake.

Finally, EnBW North America is a subsidiary of Germany's EnBW Energie Baden Würtemberg AG, a German utility and offshore wind developer that has successfully built four offshore wind projects off the coast of Germany and is developing a 900 MW project in the German North Sea that it won with a zero-subsidy bid. In 2018, EnBW entered the U.S. market and was narrowly edged out by Equinor, Shell, and Vineyard Wind in the BOEM Massachusetts auction. In fact, White told me that "we bid three times higher than any previous auction held in the United States, and we were terribly disappointed by not winning a lease area in which to build a project." He indicated the company is not dissuaded; "instead of retreating, our company has gained confidence and has decided to double down on the U.S. offshore wind market." The high-stakes game of developing the U.S. offshore wind market doesn't look like it will slow down any time soon.

TODAY'S TURBINES, TOMORROW'S PROMISE

The turbine manufacturers are gearing up to supply this offshore gold rush. GE Renewable Energy is one of them, and to get the company's perspective, I spoke with Derek Stilwell, commercial leader North America, in its Offshore Wind division. GE was a pioneer in the U.S. offshore arena, with five of its six MW Haliade turbines going into the first-of-its kind Deep-Water Wind project off Block Island, Rhode Island in 2016.

Since then the firm has scored even bigger wins. It secured an order from wind pioneer Ørsted to deploy its 12 MW Haliade machines in two offshore wind farms—the 120 MW Skipjack project off Maryland, expected to come online by the end of 2023, and Maryland's 1,100 MW

Ocean Wind project, slated for commissioning in 2024.[29] In December 2020, GE notched another victory, securing the 800 MW Vineyard Wind project.[30]

Stilwell has been in the onshore wind business for a long time, so he's seen these machines evolve to where they stand today. He commented that industry watchers have become so accustomed to the steady growth in turbine size that they now appear almost addicted to the pace of change. Indeed, many are already asking about what comes next after GE's 12 MW Haliade-X turbine (in September 2020, GE launched an uprated 13 MW version, so perhaps that addiction is justified).

In a telephone conversation, Stilwell explained the rationale behind GE's $400 million bet on the innovative Haliade platform. This was all about economies of scale, he noted. There was a clear need to build bigger turbines so that fewer of them would be required, which would improve the economics for related expenditures such as cabling on the ocean floor.[31] The challenge in getting to 12 MW, though, was that as the machines kept getting bigger, the blades also needed to be longer.

Simply increasing the rated generating capacity with larger turbines, say from 8 to 12 MW, without increasing the swept space with longer blades, implied that the economics would worsen.[32] The industry needed to add longer blades to increase energy output. At some point, Stilwell commented, "There are certain limits to loads that a platform can absorb without being significantly reinforced or redesigned."

So, GE decided to make the leap, starting from scratch and investing in the development of an entirely new platform, with a larger turbine and a larger rotor diameter with longer blades, and a step to a higher efficiency that would dramatically improve the overall economics.[33]

In one sense, Stilwell noted, the company didn't have a choice, as "there's a demand for radical cost improvements because it's a race."[34] For now, he said, the 12 MW platform is "the right turbine for the right time." He was hesitant to speculate as to where this will evolve, simply stating, "Rather than throwing a number out, I think we can say the platform has legs. We don't talk numbers, but we are already exploring the introduction of evolution to that platform. And it has already had phenomenal performance."

Stilwell pointed to the fact that even as the turbines have been continually designed to be larger and absorb more stress, the life spans have increased and are now targeted to last 25 years (that's where Yarala and the Wind Technology Testing Center come into play), "and there's always a certain amount of engineering buffer." He noted that the wind industry is just now retiring the first turbines that were installed offshore "a long, long, time ago and they had a designated life of less than 20 years, but are now well over 20 years old."

The Operational Trade-Offs

As with any machine, the expected life versus operating life is not just a function of design standards but also how operators run the turbines. If they are run harder, their life spans are shorter. It's like a car, Stilwell explained. If you have a car that can travel 60 miles per hour but the road is rough, you may have to travel more slowly than you would on a smooth freeway. Similarly, a wind turbine can have different performance based on the road it has traveled. Those machines in more stable environments can be pushed harder. To address this dynamic, GE invested heavily in modeling and creating "digital twins" that help operators know how much of the wind turbine's useful life is being consumed and whether it can be pushed harder under certain conditions. That approach ensures that the asset is not pushed past its limits, but also not underutilized.

This digital strategy is similar to that employed by GE and other operators to manage conventional gas-fired turbines. It provides improved situational awareness. Among other benefits, it can inform the user ahead of time of the optimal date for taking a machine out of service for maintenance, *before* something breaks down. That approach applies not just to the individual wind turbine, but also to the entire suite of assets. While each turbine operates in its own microenvironment, the same way each musician plays his or her own instrument in an orchestra, what each player does affects the others. Just as with an orchestra, though, each turbine can be digitally orchestrated to create the optimal output for the entire coordinated set of assets.

With wind, out-of-sync players affect the economics of the entire wind farm. Stilwell noted that it's important to operate all of the turbines in a synchronized manner: "You need to operate the whole wind farm to optimize overall production. An example of that is reducing load that turbines in the front rows can accept, and letting more wind through to the downstream machines. The trade-off for reducing flow to the leading-edge turbine results in a net energy gain."[35]

FLOATING A NEW CONCEPT

For the next few years, the offshore market will be dominated by the fixed structures attached to the ocean floor, and that's an enormous market.[36]

However, those projects are limited to coastal areas with a continental shelf. The entire U.S. West Coast, much of Europe that does not have access to the shallow North Sea, and a large portion of Asia would be unable to exploit offshore wind resources without the emergence of new technologies. That's where floating wind platforms come in. These involve putting turbines on floating platforms, such as spars tethered to the deep ocean floor.

Stilwell observed that this is "definitely a high-potential market," but "still in its infancy in terms of technology," with challenges related to tethering, anchoring, and connecting back to the mainland with cables in far deeper waters. "We see that horizon probably being closer to 2030 than to 2025," he said, but nonetheless noted that there are projects already out on the ocean delivering energy to the mainland. "There's a lot of attention being paid to it now," he said, and noted that "there are about a dozen floating foundation designers."

That aspect of the industry may well evolve before most of us know it, the same way that the fixed offshore wind industry took many people by surprise. "Part of what we continue to hear from people," Stilwell said, "is 'Did you expect this to happen and move so fast?'" In reality, the groundwork for this boom was laid over the last decade or longer. "This didn't happen overnight. This is a future that is finally here and is happening," Stilwell said enthusiastically. "It's not a question of do we have some cool turbines—yes we do—we are accelerating the development of technologies that respond to market needs, and that's very exciting."

Rahul Yarala commented during my visit to the WTTC that the next generation of offshore turbines—especially floating ones that would transmit much of the stress to the ocean rather than to fixed supporting structures—may be far larger. As a consequence, there's a good deal of interest in expanding the Charlestown facility. The challenge is how to adequately plan for a facility in a future marked by extraordinarily rapid technological progress.

The WTTC team recently submitted a white paper proposing an expansion, though Yarala said it's only conceptual at this point. The next step in creating a facility for a new generation of blades would be a detailed feasibility study to evaluate the possible evolution of the technology over the next 15 to 20 years. "Expanding the shed to 140 meters would help," he said, "but if the announcements are there for 150 meters, we would have to expand further. So, we need to spend time with developers and original equipment manufacturers (OEMs) to understand what 15–20 MW turbines mean. Floating turbines might allow them to build to 25 MW, and then we'd have to test that blade."

There's a lot at stake here, since competitors in Europe and China are not slowing down. They too are building larger machines, expanding testing facilities, and positioning themselves for the future. "It would be great," Yarala commented, "for the United States to start thinking and investing and developing strategy to be on par with Europe and China."

BUILDING THE OFFSHORE GRID

Regardless of whether floating or fixed, these turbines will need to be connected to—and integrated with—the onshore power grid. GE's Stilwell explained that it's critical to optimize the point of interconnection between the wind farms and the grid, and there is a lot of work required to integrate both the turbine technology and the grid technology, including the deployment of high-voltage lines. "I like to say that the turbines are an applied technology. They come with a certain number of options, and then there's a certain amount of defining of how that turbine is actually going to be deployed. Layout, operating, servicing strategies, grid design, and point of interconnect requirements" all matter.

Suturing these large and intermittently producing resources onto that existing onshore transmission infrastructure will not be easy. In some cases, the offshore wind farms can use the transmission capacity freed up when older power plants are retired, such as the retired Brayton Point coal-fired plant in Somerset, Massachusetts, or the B.L. England coal-fired plant and the Oyster Creek nuclear facility, both in coastal New Jersey.[37] In such cases, there's both the land for the necessary infrastructure, as well as the critical utility interconnections that would facilitate plugging into the mainland grid. The first large project, Vineyard Wind, has opted to go another route, with plans to build its own transmission and connect to the grid in Barnstable, Cape Cod.[38]

There may be another and better way to do this, though. Peter Shattuck, president of Connecticut OceanGrid at Anbaric Development Partners, explained that Anbaric has years of experience in developing underwater transmission cables, including the 65-mile-long, 660 MW high-voltage direct-current (HVDC) line running from New Jersey to Long Island, sufficient to power 600,000 homes.[39]

Anbaric's approach is to build out an ocean grid, in much the same way the $6.8 billion, 18,500 MW Texas CREZ (Competitive Renewable Energy Zone) transmission lines created a clear pathway to integrate enormous amounts of wind and other resources into the Texas power grid, helping to make Texas the fifth-largest wind energy economy in the world.[40] Shattuck suggested that the most efficient way forward would not be for each project to build its own one-off transmission system directly connected to the mainland, but instead to create an offshore network that would allow the industry to continue scaling.

Now is the time to do this, he indicated, because "we are at the first phase in the offshore wind industry, getting contracts and projects selected. Energy and supply chain, and parts of the industry, are coming over from Europe. That phase is coming to a close. The next phase is, how do you build the wind infrastructure in order to keep scaling." The problem right now is that 2,400 MW of Massachusetts and Connecticut-focused projects are going into Cape Cod, an area that has a peak demand of only 500 MW, and lacks the necessary onshore infrastructure to smoothly integrate all the energy.

The first projects will grab the easy points of interconnection, and then it gets progressively harder in an ad hoc approach. The best way to proceed, Shattuck explained, is to build the integrated transmission highway. Thus, instead of each project focusing on both generation *and* transmission, a better approach might be to create a separate focus on each area, the way Texas did with its onshore CREZ lines—which opened up a huge swathe of Texas to wind development—and the direction Europe eventually took in the North Sea as its industry matured. Opening up the transmission infrastructure to its own bidding process could also help drive down costs and help to mitigate risk in that aspect of the business. Finally, Shattuck noted, it creates a more even playing field for leaseholders further offshore, resulting in a more robust and competitive offshore industry in which developers pay a known and fixed price for transmission without even needing a long-term contract.

Conceptually, he pointed out, if done right this could be seen simply as an extension of the onshore grid, where power could be transmitted and rerouted if and when bottlenecks occur, rather than individual projects with their own generation lead lines, that are "basically a long extension cord plugging into the onshore network." Here, we can learn from the Europeans. Shattuck noted that the Belgians, Dutch, and Germans all started with transmission and then added the generation, while the United Kingdom initially began with generation and transmission included in each project. The last country is now running into siting and permitting issues, and the local regulator is opting for a transmission-first model.

The critical point, said Shattuck, is to create future optionality. In Texas, for example, they didn't build all lines immediately, but they did adopt the overall conceptual model from the outset. As Texas approached certain milestones, it built out more transmission capability, phased and expandable over time. Shattuck also noted that Anbaric is a strong proponent of high-voltage direct current (HVDC) lines, as they are more efficient with fewer line losses.

The big questions—as with any project of this scope and nature, involving multiple moving parts—are how much would it cost, and who is going to pay for it? Shattuck observed that resolving the transmission question "is going to be a massive undertaking one way or the other." Right now, he asserted, the related costs are bundled into the wind projects and typically represent 15%–30% of the total cost. "But the real wild card," he said, "is what's the cost of the onshore upgrades? For multiple projects we're talking about billions of dollars." So, if one looks at the entire undertaking holistically, the best approach is to build an ocean grid involving fewer cables in the water, which avoids onshore grid constraints and the resulting necessary investments. This would also take the burden off each wind project to build one-off extension cords to the mainland, so that cheaper energy is flowing over those lines while creating better expansion opportunities in the future.

Shattuck observed there is plenty of investment capital out there to support such a project, and summed up our conversation: "Phase 1 was, let's just get the industry up and running. Now it's phase 2, and a lot of regulators and policymakers are recognizing that if we don't think strategically, we are not going to be able to get to scale."

In November 2019, Anbaric filed an application with the Bureau of Ocean Energy Management for nonexclusive rights-of-way to develop a Southern New England OceanGrid.[41] This would be an open access system designed with the goal of facilitating development of the area's offshore wind, built in phases that could ultimately accommodate 16,000 MW of projects off southern New England.

Eric Hines summed up all of these issues in a discussion at Tufts University over lunch in early March, just before the world turned upside down. "I'm a civil engineer and structural designer. I take the long view on this thing. My job is to think about things like the 100-year storm and what's infrastructure and what's in the public interest. We are in the middle of the biggest transition in the history of the modern world and no one knows it yet. And we are talking about public infrastructure and not just private industry. Something this big, you've got to take very seriously in a way we haven't taken seriously in a very long time in terms of infrastructure build out. We're going to be inventing new ways of doing things that are in some ways old things. . . . The investments people have to make are very large and they don't pay off immediately. In a culture that values fast returns on everything, how do you make long-term investments?"

And yet, in the face of that uncertainty, the winds of change continue to blow very rapidly indeed.

10

CHARGED AND READY

In early September 2020 I received an email from a colleague, Robert Chatham, a former power plant engineer and solar energy consultant from the Bay Area in California, discussing the recent local blackouts. Chatham commented that he'd had "to deal with the PG&E Public Safety Power Shutoff outages. We have solar and a Tesla Powerwall 2, no problem for us!"

A DREADFUL LEMON SKY

I had read that Californians were likely to install as many as 50,000 residential battery installations in 2020,[1] and I wanted to find out more about that experience. So I quickly set up a teleconference call and reached Chatham on September 9. That turned out to be a somewhat apocalyptic type of a day in San Rafael, where he lived.

As we spoke, Chatham held up his computer camera so I could see the online map of local utility Pacific Gas and Electric Company's "Public Safety Power Shutoff" map. This is the map PG&E uses to inform customers in real time where "if high temperatures, extreme dryness and record-high winds threaten the electric system, it may be necessary for us to turn off electricity in the interest of safety." The utility has come under tremendous pressure for its failed equipment having started deadly fires in previous years. In 2020, it pleaded guilty to 84 counts of manslaughter for its role in starting the deadly Camp Fire blaze that incinerated the town of Paradise.[2]

That fire led to lawsuits, an eventual bankruptcy filing, and an increasing reliance on prophylactic power outages, de-energizing lines so they cannot create the sparks that set the state ablaze. On the map Chatham held up for me to see, purple inverted triangles dominated, highlighting the current outages. Other symbols denoted the number of individuals affected at any location, with some places up to 4,999. Of the 167,000 homes without power on September 9, Chatham's town of San Rafael hadn't yet been affected, but he said that could only be a matter of time; "The power runs north to south, so if a fire starts to the north up in wine country, they shut us down," he explained.[3]

Then he took his computer outside to show me the view of the San Rafael skyline. A muted reddish pall hung over the land, and the visibility to the Bay was limited. The sun's dim light was completely obscured by clouds, with ash falling from the sky. "Get a load of this," he commented, "the wildfires have changed our landscape totally. Today it's like being on Mars. The camera doesn't show the true redness you actually can see. This isn't like being in China, where you have really bad air pollution. No, this is like being on Mars. It's truly weird. It's only 60 degrees. Here it is 11 o'clock in the morning. The sun is being blocked out, and the ash is about 3,000 feet and above. It's everywhere. But there's no smell."

Chatham walked with his camera up the slope to demonstrate the solar panels he had installed on his roof two years ago to cut his power bill. As we spoke, he traced his index finger down one of the panels, holding it up to the camera so I could see the accumulated ash on his fingertip. The fires were not close (about 150 miles to the north), he said, but they were everywhere, up north in Oregon and spread across parts of his own state.

Chatham was blessed with both luck and foresight. He and his wife moved into the house in November 2017, and one of the first things they did was invest in efficient lighting, swapping out the older lights for high-performance LEDs. Then in April 2019, once they had gained some familiarity with the house and its energy consumption patterns, they put 18 solar panels on the roof, totaling 6.4 kW.

A year later in April 2020, they took the plunge and added a Tesla Powerwall battery system and a 7.6 kW inverter from Solar Edge. He took me virtually into his garage and showed me the inverter and Powerwall system neatly attached to the wall. "We haven't had an electric bill for a year" he said. Then he paused and corrected himself. "Well, we had one for $10 once as a result of charging the Tesla Model 3." Of course it helps that Chatham doesn't have air conditioning or an energy-hungry pool pump. That cuts his demand considerably.

His Tesla app shows how much power flows from the solar array to the battery, and from the battery to house. "On a day like today, I keep the battery fully charged. For us, as long as we cut back on outdoor lights and things like that, it's power all the time, as long as the sun comes up in the morning. We're off the grid basically."

Those sporadic power outages used to pose a problem, but not anymore. With the solar and the batteries, Chatham and his wife are now immune. "We don't worry about the lights. We don't worry about what PG&E does, we are just so independent." As long as you can afford it, he says, "It adds value to the home and with a battery it's a nine-year payback." That value proposition is catching on. At last count, three of his neighbors had recently signed up for solar and battery combinations, buying their way toward energy independence and freedom of mind.

IT'S NOT JUST THE RESIDENTIAL SECTOR: SOON BATTERIES WILL BE ALL ACROSS THE GRID

Chatham's experience with batteries, and his ability to store energy for later use, is emblematic of the change that is taking place with energy storage across the electric grid. In the recent past, electric energy was stored upstream (sometimes literally, in the case of hydropower) of the electric generating plant. Whether in the form of raw fuel in a coal pile, natural gas in a pipeline, or water behind a dam, all of these energy sources were eventually converted into electrical energy when that power plant was called into action and dispatched.

In today's world, increasing levels of economical wind and solar energy permeate the system. However, the "fuel" only shows up when the wind blows and the sun shines, so grid operators have increasingly less control of when and where the electrons are generated. Weather now affects both sides of the electricity equation, with marginal electricity demand driven by temperature and marginal energy supply increasingly affected by the available wind and sunshine.

The challenge for grid operators is to take this growing quantity of variable renewable energy when it's generated and determine how best to integrate and optimize it. Gas-fired generators can absorb some of that stress, ramping up and down frequently to accommodate intermittency. However, that increases wear and tear on the machines, and they can only do so much as the renewable resources contribute more and more energy to the grid.

The other thing gas turbines and other grid resources cannot do is absorb energy, which is sometimes just as critical as being able to release it. As a consequence, energy storage—and especially the growing fleet of battery storage facilities—is rapidly becoming the critical ingredient in allowing the grid to accommodate more renewables. Storage—especially in the form of batteries—can be deployed across the entire grid. It can be implemented in the form of massive utility-scale projects to support renewables, or help shore up the transmission and distribution infrastructure. It can reduce or eliminate the need to build new substations or power plants in or near cities. And it can help customers avoid high costs on their electric bills, or supply energy during blackouts—keeping people safe, comfortable, and healthy. Ultimately, storage creates the potential to move the electricity grid from a just-in-time instantaneous electricity generation and delivery system to something far more flexible, resilient, and economically efficient. It's increasingly becoming the glue that will help hold everything together and allow the grid to operate more efficiently.

In a January 2020 conversation on the topic of batteries and storage, NEC Energy Solutions (NEC) CEO Steve Fludder had conceptualized the utilization of batteries on the power grid in a way that made sense to me.[4] The entire evolution of the grid to a lower-carbon, renewables-dominated system, he had noted, was really a shift in how and where the electric energy is stored, from being contained in the raw fuel to the physical storage of electrons in batteries within the grid itself. It was a migration toward a "digitally optimized" system, he said, in the form of batteries and AI-infused software applications that manage them.

At that time, NEC Energy Solutions was one of the larger vendors in the storage space. A subsidiary of the Japanese electronics and information technology giant NEC, it had close to 1,000 MW of battery storage in the field or in the development pipeline. I wanted to delve deeper into the topic, so on March 3, 2020, I headed out to Westborough, Massachusetts, to visit and meet the leadership team in person. I entered through the glass doors of the facility and presented myself at the front desk. Here I encountered the first tangible indicator of the invisible storm from abroad that was about to sweep the United States and wreck our lives and our economies, and I completely missed its significance at the time.

The receptionist requested that I sign an attestation that I had not traveled outside the United States within the past 30 days. And what would happen, I jokingly asked, if I had indeed been outside the country? She looked at me with all seriousness, and replied, "Then you would not be let in." I should have gone home that day and liquidated my entire 401K, but like almost everybody else, I missed the sign that this deadly Black Swan was headed our way. It was a Black Swan that would also have a hand in leading to the demise of the company I was visiting, just three months later.[5]

My host, Chris Lawson, NEC's director of marketing communications, showed up and warmly invited me in. We grabbed a cup of coffee and headed to the conference room, where I met part of the team that led NEC's storage business.

CONSTANTLY EVOLVING BATTERY TECHNOLOGIES

Assembled in the conference room were Michael Hoff, vice president of research and technology and CTO; Roger Lin, vice president of marketing; and Ramdas Rao, vice president of software engineering (CEO Fludder could not make it that day), and they took pains to explain the rapidly evolving industry to me over the course of several hours.

The vast majority of lithium-ion battery technologies being deployed on the power grid are the same ones being used in electric vehicle (EV) industry.[6] The stationary storage industry today is the tick that rides on the EV dog; wherever that dog wants to go, the grid storage industry is along for the ride. That dynamic will likely change as the stationary storage industry grows in its own right, but for now, the new storage projects being installed across the globe largely rely on the same batteries used in the global EV industry.

Like most storage developers (with the exception of Tesla, which uses its own product), NEC sourced its batteries from various vendors to avoid the risk of being tied to a single supplier. Its projects use the nickel-manganese-cobalt (NMC) technology, the predominant chemistry used in stationary storage in 2020.[7] The majority of vendors utilize one of two major lithium-based chemistries—either NMC or lithium-iron-phosphate (LFP). The storage industry as a whole is evolving toward the LFP chemistry since it has no cobalt (which is both expensive and associated with child labor in Africa) and is more stable.

The sheer mass of the growing industry is worth contemplating. Tesla's battery "Gigafactory" in Nevada, for example, encompassed approximately 5.3 million square feet (roughly 1.9 square miles) of operational floor space as of 2020 and that size represented only 30% of its ultimate planned footprint.[8] However, Tesla is not the only company with such massive ambitions. In September 2020, French hydrocarbon giant Total announced plans for two gigafactories in Europe that together would exceed Tesla's facility in total output.[9]

Meanwhile, China's Contemporary Amperex Technologies (CATL) is planning on a factory in Germany that could ultimately be three times larger than Tesla's.[10] CATL boasts as many as 1,000 scientists with advanced degrees in materials science, and broke ground on a 44-acre research campus in June 2020 to investigate and develop the next generation of battery technologies.[11] It estimates an additional 1,000 researchers will be employed in this effort.[12]

That's the other critical fact defining the battery industry: the underlying technologies are still very immature, with enormous opportunities for technical improvements (just the sort of thing supercomputers can help with). CATL's new research facility (and those of competitors—nearly every major company is investigating what's next) will shake up the space in the years to come as it investigates many new promising technologies, including lithium-metal, solid-state, and sodium-ion batteries.[13]

CATL supplies batteries to the likes of Volkswagen and others. It even supplies Tesla for its market in China, and the two companies reportedly worked together to develop a "million-mile battery" to be released in China in late 2020 or early 2021.[14] In June 2020, CATL's CEO announced a battery that would last up to 16 years and offer a car up to 1.24 million miles of range (the equivalent of five trips to the moon, for a driver so inclined). That compares with today's EV battery warrantees that typically cover 150,000 miles (one-eighth of the new battery) or eight years.[15]

Such longer-life batteries have the promise to radically remake both the transportation space and the electric grid. Until now, cost-effective lithium-ion batteries have been fundamentally limited by their inability to cycle for more than roughly 2,000 cycles (a complete process of charging and discharging) before the battery is spent.[16] One million miles is great news for drivers. It means the chassis will never outlast the battery and, in theory, the battery could potentially be pulled out and put into a new car. Quite possibly, that used and fully paid-off battery could make its way onto the power grid, serving the remainder of its years supporting renewable energy installations and stabilizing the grid power supply.

Hoff's technology team at NEC was aware of such ongoing developments, and was constantly on the lookout for "anything that would help improve our competitiveness out in the future in different customer applications." Such a level of scrutiny is critical in the global market place, as tens of billions of dollars are being invested globally, and technological obsolescence is a constant concern.[17]

In our March meeting, Lin commented on the rapid pace of that change: "I speak to a lot of analysts to see what they are seeing. It's tough. You do a deep dive, and then it's out of date." In fact, battery technologies are changing so rapidly, it's likely that within a few years the iron-based, million-mile battery or some variant thereof will be the dominant chemistry for stationary applications—at least for a while.[18]

IT'S ULTIMATELY THE SOFTWARE THAT MATTERS

These lithium-ion batteries in our cars and on the electric grid today are wondrous devices that absorb, store, and release energy. However, they also have some fundamental characteristics and limitations that affect their performance. They don't like to operate in environments that are too hot or too cold. They don't like to be charged too quickly, or beyond certain levels. The final 20% of the charging cycle (from 80% to 100% full) can take roughly half of the charging time. And the battery's performance when it has roughly 20% of the energy remaining in the battery is fairly poor and drops off quickly. Further, if lithium-ion batteries become too hot—usually the result of a short-circuit—they can overheat, leading to a process known as "thermal runaway," where the heat from one failed battery propagates to the next and causes a chain reaction that may eventually result in a fire.

The battery itself is also a passive device and essentially a dumb brick if you don't apply intelligence to it. If somebody or something is not telling that battery how much power to absorb or release at any given moment, and for how long, it has no intrinsic value. For the storage company operators, having the right data and being able to act on it is the critical piece of the puzzle. NEC, for example, collected terabytes of data annually for safety and reliability purposes, but also for operational issues and to optimize performance (among other things, battery cells age as they cycle, so operators need to know when to swap out assets as they near end of life).

That's where the software engineering teams come into play. At NEC, Rao's group was responsible for the controls managing each battery system at the individual site level, with software remotely monitoring the devices, pulling data up into the cloud where it can be analyzed for trends, and looking for any defective components or weaknesses.

That software was powerful, applying a neural networking approach that could predict the life of a whole fleet of energy storage devices in 10,000 homes in Japan, using data collected every five minutes over more than four years. Neural networking allowed NEC to examine correlative relationships and identify patterns that could predict future behaviors. Based on that information, the team could determine the "fade curves" of the batteries and know exactly how fast they were wearing out. This in turn allowed the team to determine the benefits and costs of every market strategy, and indeed how much cycle life would be consumed during each application.

BUSINESS CASES AND
APPLICATIONS ARE CONSTANTLY CHANGING

It's not only the batteries that are rapidly morphing. As the grid quickly evolves, the use cases—how and where batteries can be deployed to make a profit—do as well. In many cases, operators resort to the practice of "value stacking," utilizing the battery like a Swiss army knife to perform different functions at different times. For example, a battery on the Texas grid might charge at night using low-cost wind power, and release more expensive electricity back into the grid during the hot afternoon hours when prices are high—as high as $9,000 per MWh. By contrast, a similar battery in solar-soaked California would typically charge with cheap mid-day solar energy and sell that energy into the grid during the evening peak demand period.

In either location, at other times those batteries might also balance the desired grid frequency at 60 Hz, instantaneously absorbing or releasing power in response to a signal from the grid operator in order to maintain grid stability. Or they might provide needed capacity to the grid for reliability purposes when demand is extra high.

There are multiple ways for battery owners to generate revenue, and the NEC team indicated that it was normally seeing two or three revenue streams for any given project, and had identified 18 different operating modes in which its batteries could make money around the world. Some of these involved short-term and larger releases of electricity (power) and some were longer-duration (energy).

However, electricity markets change constantly these days. The addition of more renewable assets on the grid may rapidly and dramatically increase the value of a storage project, since its ability to balance the grid will become more valuable. By contrast, the arrival of other storage projects in the same neighborhood, or a change in the market rules, can significantly hurt the economics. In some cases, the market for battery services change so quickly that the battery may be best applied somewhere else where the economics are more favorable. The NEC team cited the case of one battery installation deployed in Huntington Beach, California in 2008. Three years later, the business case for storage was far more compelling on the East Coast, so the company moved the battery across the country to just outside of Philadelphia, where it served for seven or eight years. Then, after Hurricane Maria hit in 2017 and Puerto Rico's grid was in disarray and needed support, the unit was shipped, yet again, to the island.[19]

One thing you can't do as an operator, though, is use multiple applications of that Swiss army knife that may put the grid at risk. If you're contracted to sell reliability services to the grid to help provide critical services to avoid a potential blackout, for example, then your battery had better be fully charged when it is called upon. If you've already used it to take advantage of market opportunities and it's depleted and unavailable, that puts the grid at risk. And if your battery is not available, the resulting penalties can be significant.

A THIRD ASSET CLASS FOR THE GRID

In our conversation, Lin commented, "Storage is on the shelf for the toolkits of grid planners. . . . With coal retiring, and gas falling out of favor, we will see more wind and solar, which makes operating the grid that much tougher. . . . Batteries will be that third asset class. You have generation, transmission, and now you have storage." The implications of that simple statement are enormous, and batteries could change a lot of about the way we produce, deliver, and consume electricity.

By far the largest application of battery storage at present is to "firm up" intermittent utility-scale renewables projects. If clouds cover a solar array, for example, batteries can release energy and pick up the slack. Alternatively, a standard package of four hours' worth of batteries can be used in a sunny market like Arizona, California, or Hawaii to absorb and shift that solar output from mid-day, when there may be a surplus of solar energy, to the late afternoon and evening demand peak when it is far more valuable. That's what solar developer 8minute is planning with its proposed 24,000 MWh of batteries to back up its solar projects in California, Texas, and areas of the Southwestern United States.[20]

In the "bulk power" generating system, batteries have the potential to take over the role of gas-fired peaking generation plants. These are only called on to meet peak demand a limited number of times per year and often run only a few hours each time they are dispatched. Batteries have the potential to perform this service more cleanly and economically, and these peaking projects can be as large as the power plants they are displacing. One recently announced permitted potential project may involve 1,500 MW and 6,000 MWh of batteries—roughly three times the energy of all the batteries installed in the United States in 2019.[21]

On the distribution grid, batteries can be used to help minimize the need to build out new infrastructure. Multiple storage projects are being developed from Cape Cod to Arizona, typically where long distribution lines have been used to serve small populations.

And from a customer perspective, the value of batteries is not limited to the residential sector and the Robert Chathams of the world. Batteries increasingly make sense in the commercial sector as well. Installations might be for resilience, or simply to help facility owners save money. In New York City, for example, Glenwood Management began deploying advanced lead acid batteries in its buildings in 2012, eventually building out its portfolio to 15 buildings.[22]

Joshua London, Glenwood's vice president of management, told me he was skeptical at the outset. His team first met with storage company Demand Energy in 2009, and he told me his first reaction was, "Oh, come on, we are going to stick these giant batteries in, like toys? But then we thought 'this could actually work,' and in 2011 we did a small 10-kilowatt system to see how well the process would go and what it could do. It worked. And then as years passed, we did other projects and it became part of the vernacular."

The company recouped its initial investment in three and a half to four years and now saves about 15% on its utility costs from its batteries. London indicated that the units, which average 100 kilowatts in size with four hours of duration, are typically placed in the parking garages under the buildings and take about two parking spaces.

MICROGRIDS

Batteries are also increasingly being integrated into miniature grids that can be disconnected or "islanded" from the larger power grid in the event of a power outage.[23] These microgrids can be highly complex, balancing supply and demand on a campus such as a university or a military base. They can take the form of the sophisticated microgrid at the Parris Island Marine Corps Recruit Depot in South Carolina, which calls on 20,000 solar panels, 8 MWh of batteries, a highly efficient combined heat and power plant (that utilizes the waste heat), and three diesel generators.[24] Or they can be less complex, such as the 110 Red Cross microgrids in Puerto Rican schools, each equipped with a 5.7 MW solar array and 11 MWh of batteries, allowing the schools to serve as emergency shelters.[25]

SAFETY IS CRITICAL

A lot of energy is held in a battery storage unit, so not surprisingly, safety is a major issue. NEC's Lin commented that ensuring safe battery performance was like keeping a car on the road. In much the same way cars have adaptive cruise control and lane departure technology, NEC used algorithms and constantly monitored batteries in operation—especially voltage and temperature—to ascertain whether the battery was operating within safe and allowable parameters. "If they do go outside those limits," he said, "we will shut stuff down to keep that car from going off the road. Even if the car is driving between the lines, we will use monitoring algorithms to see if the car is weaving inside the lanes. The battery could be within safe allowable tolerances, but behaving differently, so we can go check it out."

All of that intelligence was built into NEC's command platform, which oversaw multiple installations including batteries from multiple companies, and occasionally that level of diligence would indeed identify a critical issue. In one instance, the team identified abnormalities in a system operating in New York, "so we pulled it and we found it had swollen cells inside. Left unchecked, it could have led to an unsafe situation. It's a way to keep an eye on things."

Like other competitors in the industry, NEC's group was intensely focused on the safety issue: one bad experience could not only put somebody in harm's way, but also create a huge damper on the industry's growth prospects. To that end, the group was awaiting the results of the investigation into a 2019 incident at Arizona Public Service (APS), where firefighters had been critically injured responding to a thermal runaway event in an energy storage installation.[26]

Here, too, the NEC team indicated, it was all about effectively making use of the relevant information and providing appropriate "situational awareness." Rao pulled up a screen in the conference room and demonstrated a simple app the group had created. It included project location, address, and the basic information first responders would need to know.

Lin said, "With any phone you can log into the URL and look inside the battery container and ask, 'what are the conditions inside?' It provides you with information on state of charge and temperature. What's it doing right now? Has the fire suppression system gone off? Is it normal inside? It's a way to provide better situational awareness to first responders safely and remotely—looking into the container without going into the container."

The challenge with lithium-ion batteries is that if a "thermal runaway" situation occurs inside the container, there's a buildup of both heat and gases released as the battery temperatures soar. In the case of the Arizona incident, a faulty cell apparently short-circuited and the heat propagated to the next cell, then to the cell beyond that one, and so on. As the reaction intensified, the batteries got hot enough to melt the aluminum housing, and the temperatures in the container were sufficient to melt and fuse some of the modules together. The fire suppression system went off in response to the incident but may have actually served to increase the concentration of gases, which were not ventilated. All that was needed to ignite the mixture was oxygen, so when the firefighters opened the door, it was like a bomb,[27] with one firefighter thrown 73 feet through the air and badly injured as a result.[28]

While the safety issues related to lithium-ion batteries are serious, they are generally manageable, and many of the storage companies have made improvements since the Arizona incident. Among other things, companies have developed improved layout and spacing of batteries within their purpose-built containers to minimize the potential for thermal runaway. Many have also created better approaches to detecting and venting gases to remove the potential for a dangerous explosion. Some companies are even deploying "deflagration panels" that would direct the force of the explosion vertically in order to minimize the threat to humans, and others are modifying the actual containers themselves.[29]

THE POTENTIAL FOR
USED ELECTRIC VEHICLE BATTERIES

While huge volumes of new batteries are being expressly deployed in the electric grid, many industry analysts are convinced that second-life EV batteries will also prove to be an enormous and cost-effective resource, flooding into energy markets roughly a decade after EVs were initially sold.[30]

In fact, there are already storage projects all over the world deploying used EV batteries, including a soccer stadium in Amsterdam that combines new (85) and used (63) Nissan Leaf battery packs to support solar energy on the roof, reduce diesel needed for backup power, and sell services back to the power grid.[31] Standards lab UL has also developed its "Standard for Evaluation for Repurposing Batteries," and in 2019 certified a joint venture between Nissan and Sumitomo, including ratings to identify the battery's health and validate its capability to perform in second-life applications.[32]

There may be potential there, but NEC's Hoff was not so sure the growth of the second-life storage industry will be an inevitable outcome.

"A lot of people think, 'I have 80% life left in the battery when its capacity is down to 80% of its new condition' in the EV. That's completely false," he asserted. "You probably only have 10% or 20% left, if that." Hoff noted that many major manufacturers like LG and Samsung don't guarantee batteries below 60% and sometimes 70% of remaining life, because beyond that point there are numerous uncertainties. "You don't know where it's been, or how it's been used. Has it been in the desert or the Arctic? Has it been charged on super-fast chargers twice a day?"

Depending on how it has been deployed, each battery may function quite differently in a second-life application, creating some significant operating challenges. Hoff paints an amusing picture to illustrate the reason for his skepticism: "You are going to have operators on roller skates in warehouses going after red lights, and trying to keep the system operating and available. The guys running this thing would love their jobs—they would have to. The battery is free. That's the only redeeming factor about this."

Hoff warned, "The probability of a battery failure is higher because the battery has been through most of its life. . . . Batteries for cars are intended for low cycle life and long-distance travel and to get the car through the warranty period. . . . The battery enjoyed a cozy life in a car and now it's going to a labor camp in Siberia. In addition, manufacturers will not support a formal official communications link between their batteries and any other third-party system. All of the communications with these packs will be built from scratch unless the original manufacturer is somehow deeply involved with the effort." Hoff didn't say it was impossible, but thought it was going to pose some significant challenges.

THE INEVITABLE NEED FOR LONGER DURATION STORAGE ASSETS

The pressure to decarbonize the electricity grid has grown significantly in recent years as the mounting evidence continues to demonstrate that climate change is going to have serious effects on the planet. Fortunately, at the same time that this pressure increases, the costs associated with wind and solar projects continue to fall roughly 70% over the past decade for wind and 89% for solar energy.[33] In many cases, these are now *the most cost-effective new resources to add to the grid*. However, as discussed earlier, the output of these resources is highly variable, both from second to second and minute to minute (think cloud cover and wind gusts) as well as hour to hour, day to day, and seasonally as well. Other resources are needed to back up these resources in order to provide the predictable flow of energy that can match societal demand (itself highly variable and somewhat weather dependent).

Batteries can help electricity grids address the short-term intermittency issues associated with renewable energy, and even cost-effectively shift output for as long as four or more hours. In fact, a recent study from California's three utilities demonstrated that when solar energy was combined with four hours of storage, supply was matched to peak demand on the grid over 99% of the time (without batteries, that same resource only lined up with peak demand 6% of the time).[34] But what happens over longer periods when the wind doesn't show up, or the sun doesn't shine?

Today, pumped hydro storage can offer that longer duration service, and, in fact, that technology is currently the largest storage resource on the grid, representing about 95% of all utility-scale storage in the United States.[35] With pumped storage technology, water is pumped up inclines into reservoirs during low-cost periods and released when energy is more expensive. There are currently 42 such projects in the United States, with numerous additional projects in the preliminary permitting stage.[36] In all likelihood, though, few of them may ever be built. That's because these projects are very expensive (costing in the billions of dollars in many instances), geographically dependent, take many years to build, and are increasingly difficult to site for environmental reasons.

The ability to create more flexible and widespread long-duration energy may eventually be addressed by the production and storage of "green hydrogen" produced from renewables. In the meantime, though, some astonishingly creative long-duration storage technologies have recently emerged. These include cryogenic storage from Highview Power—a fancy word for cooling and liquefying ambient air to −320° F and storing it in low-pressure insulated tanks (which are already being used in various industries). The air is then warmed up, causing it to expand 700-fold and creating pressure that spins a turbine, generating power. The heat used during compression can be utilized to warm up the liquid gas, while the cold temps generated though the expansion process can be applied to cooling the air upfront. Round-trip efficiencies—the amount of energy retained throughout the process—range between 60% and 75%.[37]

Then there's compressed air energy storage (CAES). To date, there are only two CAES projects in the world—one in Germany and one in Alabama. CAES uses cheap off-peak electricity when demand is low (or, in the future, when renewable generating output is high) to run a compressor that stores air under high pressure—between 650 and 1,058 pounds per square inch in a confined salt cavern (roughly 10% of the pressure in a high-pressure oil well). The air is then released from the cavern to a local gas-fired turbine to generate electricity. Gas-fired plants need compressed air, mixed with natural gas, to run the turbines efficiently, but compressing that air may require more than half of the plant's energy—which is fed back into the system to the compressor.[38] Traditional CAES represents energy storage, but in this model must be paired in tandem with a power plant to operate.

There is, however, a new CAES model that may hold some promise. Startup Hydrostor is working on the deployment of purpose-built, hard-rock storage caves to store compressed air and a compensating water reservoir that maintains the system at a constant pressure. Unlike the projects described above, which rely on the existing geology, these systems can be sited where the grid requires them.[39] The first of these projects is already active in Canada, with additional projects in the pipeline for Canada, the United States, Chile, and Australia.

Another startup, Energy Vault, uses a gravity-based storage system involving cranes on towers as tall as 500 feet—the height of a 35-story office tower. These cranes will raise and lower huge 35-ton composite bricks, storing and releasing energy in the process. These towers can be developed in multiple configurations, offering 20–80 MWh of quick-release storage capacity and 4–8 MW of power for as long as 8 to 16 hours. Only 10%–20% of the energy is lost in each round trip.[40]

There is also a project, slated for Nevada, involving specially designed and very heavy trains that would move up and down a mile-long elevated track, absorbing energy on the ascent and releasing energy during its descent, and offering fast-response grid balancing services into the power grid. This Advanced Rail Energy Storage project has been permitted but is not yet in construction.

Then there's the new mysterious technology from Form Energy, which promises to offer 150 hours of storage duration.[41] The firm uses an aqueous air battery system about which little is known, and has signed a contract with Minnesota utility Great River Energy for a 1 MW, 150 MWh system.

Finally there's also green hydrogen, which may hold great promise as a long-term storage medium. Green hydrogen can be created by using renewable electricity to power electrolyzers that separate water into hydrogen and oxygen. Hydrogen can then be transported and stored, and used to generate electricity at a later date, either through specially adapted power plants or by utilizing fuel cells. The critical challenge here will be to reduce costs and improve system efficiencies so that green hydrogen is cost competitive. In the past few years, utilities, large energy companies, and governments around the world have announced future investments totaling in the many billions of dollars in green hydrogen. While there have been many announcements, no projects of scale have yet been built, leaving many uncertainties as to whether this approach will live up to its anticipated potential.

It's still somewhat unclear who will win the race to develop an affordable and broadly applicable long-duration storage solution. What is indisputable, though, is that there will be many improved battery chemistries and other storage technologies developed in the years to come. The prize is too big to ignore, as the ability to capture and store electricity for longer periods becomes increasingly valuable. Energy storage is rapidly becoming that third asset class, enhancing clean mobility, creating more independence and resilience, and accelerating our efforts to liberate the grid from its century-old dependency on hydrocarbons.

11

TEACHING AN OLD DOG

Redefining the Role of the Gas Turbine

When I received an invitation from GE's gas team to visit them for a full-day background discussion about all things related to gas turbines, I jumped at it. Gas turbines are large engines (similar to those on an airplane) that typically burn natural gas or liquid fuels at high temperatures and pressures to create electricity. They are now the leading generator technology in U.S. power markets, so I wanted to understand the technology better.[1] I drove from Boston to the GE complex in Schenectady, New York, and on a preternaturally warm day in early March, navigated my way past yet another guard shack to the visitors' center, where I met Jim Donohue, GE's executive marketing director. Before we entered the building to meet the team, Jim explained that when Thomas Edison founded General Electric in the late 1800s, he chose Schenectady to be its headquarters, largely because it sat perched on the edge of the Mohawk River, with direct access to the Erie Canal. During its heyday in the post–World War II era, it was home to over 50,000 professionals. Since that time, many of the buildings have been torn down, but the location remains central to GE's turbine business, and its steam turbines and generators are still made here.

We entered the building and took the elevator upstairs, where I was ushered into a room with a team of around 10 professionals, either physically present or calling in remotely. As they introduced themselves, I was struck by that fact that many of them had 20 or even 30 or more years in the turbine business. Brian Gutknecht, marketing leader at GE Gas Power, led the discussion and fired up the slide deck that would inform the discussion for the next seven hours. We discussed everything from the operating regimes of various machines, to their evolving role in power markets, to the potential for hydrogen and carbon capture. A full day of involved discussion from a highly educated and passionate team left my head spinning, and my notebook full.

THE STEAM TURBINE FACTORY

Before I left the vast complex at the close of the day, Donohue took me over to Building 273 to see the factory floor where the steam turbine rotors are built. The lobby suggests your typical office building, with frosted-glass entrance doors sporting circular geometric turbine designs. That lobby completely belies both the size of the facility and the level of activity inside. In the elevator on the way up to the third-floor viewing platform, Donohue commented that it would be just our luck if the crane was in the way. I was thinking, "Well how bad could that be?" As the door opened, I found out. A massive yellow 100-ton crane supported by stanchions on both sides of the enormous bay blocked our view almost completely.

So, Donohue suggested we walk the floor (even better) of the quarter-mile-long building. We took the elevator back down to the first floor, donned safety glasses, and headed out to the factory floor. As we toured the building, Donohue explained what we were looking at. Our visit was between shifts, so the place was not humming with activity, but that also made it easier to hear Donohue's explanation as we traversed a path between clearly delineated yellow lines on the gray concrete floor that I was not to cross. The smell of machine oil pervaded the entire space. He pointed out that certain sections of the floor were covered by wood, with the cut ends facing up, darkened by years of oil and grit. Wood would absorb the force of anything that fell, he explained, greatly decreasing the likelihood of breakage. The floor was also crisscrossed by steel rails that looked like train tracks. And in fact, they were. The large rotor blanks come in from Japan by rail from the port in Albany on the Hudson River, 20 miles away, with the tracks leading straight into the factory. Further down the line, a small locomotive and a rail car were visible on the factory floor.

Overhead, numerous cranes—some up to 300 tons—perched on supports that ran lengthwise down the huge industrial shed. Up and down the building, enormous industrial machines were attached to roughly 30-foot, gleaming steel rotors in the making, ready to resume their tasks of cutting and machining once the shift began. These massive rotors were being machined and outfitted with blades that would ultimately translate the heat and pressure of the hot steam into kinetic energy and spin inside of the turbine, generating electricity.

The turbine factory floor. *General Electric*

THE ASCENDANCE OF
GAS TURBINES IN THE UNITED STATES

In the late 1990s, during the dawn of electricity competition, U.S. utilities were forced to sell off much of their generation fleets to merchant power companies. These new merchant entities were created to sell power into the markets, operating as efficiently as possible and maximizing profits. They usually had a trading arm, and many were offshoots of existing utilities or large energy holding companies. Anticipating a heyday in the growth of the overall power sector, a half dozen or more of these entities (AES, Aquila, Calpine, Dynegy, El Paso, Mirant, and Williams were just some of the names) sprang up in just a few years. Many of them set off on a capacity-building spree, and nearly 100% of that new generation fleet relied on a single technology: the combined-cycle gas-fired turbine, or CCGT.

In the late 1990s and early 2000s, natural gas–combined cycle units quickly became the dominant additions to the generating fleet. Relatively efficient, cost-effective, and with construction lead times that were at least twice as fast as coal or nuclear plants, CCGTs rapidly emerged as the default supply option. In just the five years between 2000 and 2005, the industry went on an unparalleled expansion, installing as much as 55,000 MW of capacity in a single year, and over 185,000 MW during the period.[2] To

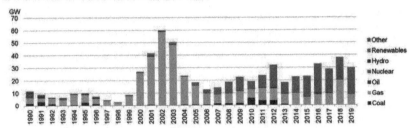

Annual U.S. generating capacity additions. *Courtesy of "Sustainable Energy in America Factbook," by BloombergNEF and produced in partnership with the Business Council for Sustainable Energy (2020),* https://www.bcse.org/factbook/#

put that in perspective, the country has roughly 1.2 million MW of total generating capacity.[3] Looked at another way, gas used for electric power generation has increased 60% between 2005 and 2019.[4]

In the following years the amount of new gas-fired generation abated, but the industry still added roughly 10,000 MW of new plant capacity annually. At the same time, the fleet of coal-fired and nuclear power plants were beginning to age out, but they were still holding on and running at relatively high utilization factors until an unexpected revolution occurred that shook the entire energy marketplace. That was the arrival of hydraulic fracturing, or simply "fracking."

FRACKING CHANGED EVERYTHING

Fracking, the commercial exploitation of U.S. shale reservoirs, involves technology, heavy machinery, water and land issues, property rights, money, and the law—in other words, it's a perfect American story.

Well below the surface in states such as Pennsylvania, Ohio, Texas, and the Dakotas lie massive sedimentary shale basins, formed from ancient seabeds over 300 million years old. These hydrocarbon-rich reservoirs have been known to geologists for many decades. However, it has only been in the last decade or so that the technology and know-how has been developed and scaled to tap the oil and natural gas trapped within these horizontal shale layers and flood the market with new supplies.

Fracking involves taking two separate technologies—horizontal drilling and hydraulic fracturing—and combining them in a new and highly effective approach. For more than a century, conventional oil and gas wells have drilled vertically into the earth's surface to tap specific concentrated deposits of oil or gas trapped in reservoirs that were sealed by impermeable rock above. Those reservoirs were often marked by high pressures and high permeability (we have all seen the old pictures of "gushers" shooting geysers of oil into the sky—the same high pressures that led to that to the 2010 DeepWater Horizon blowout in the Gulf of Mexico). With the shales, it's different. The "tight" oil and gas locked within the shale is under pressure, but doesn't move easily through the pores of the vast horizontal shale formations.

That's where horizontal drilling comes into play. This approach involves drilling vertically into the earth, as deep as a mile or two in many cases. Operators then rotate the drill rig 90 degrees to extend horizontally into the layers of rock where the hydrocarbons exist trapped within the shale formations.

Before the utilization of horizontal drilling, drillers couldn't access the shale layers. These geological structures are analogous to a layer cake, with a line of frosting between each of the sections of the cake. Drilling vertically would only provide access to a small section of the frosting. But by going sideways, one would be able to access huge amounts of frosting, and do it quite cost-effectively.

These lateral drilling extensions can extend over 10,000 feet. And, aided by highly advanced rigs and drill bits, these operations are amazingly efficient at punching holes in our planet. Some can move sideways at nearly a mile per day, and average wells in North Dakota's Bakken formation can frequently travel 10,000 feet down and another 10,000 feet laterally in two weeks to a month.[5]

But that ingenuity alone won't bring success in coaxing the hydrocarbons to flow up the well. No, for that you need more brute force, coupled with enormous quantities of water and sand. The second part of fracking involves pumping millions of gallons of pressurized water, mixed with chemicals[6] and sand, into the hole. The pressurized mixture—as high as 15,000 (psi)[7]—fractures the rock in multiple zones along the drill pathway. The sand then lodges in the newly fractured rock, keeping the pores of the rock open, while the chemicals improve the flow of the hydrocarbons out of the rock. If the vertical pipe is properly cemented and sealed, it passes through the shallow aquifers that we rely on for drinking water and heads right down to the pay zone. However, if there is a fault in the process, groundwater contamination may result, and that has occurred in various poorly completed wells.

Perhaps of greater concern, though, is the water-intensive nature of the hydro-fracking process. According to the U.S. Geological Service, fracking can consume between about 1.5 million to as much as 16 million gallons per well.[8] Multiply that by hundreds or thousands of wells, and water use becomes quite significant. Nationally, one estimate for 2020 water use is over 225 billion gallons.[9] The lion's share of the water that is pumped down into the earth remains in the formations, but approximately 25% of the contaminated water is recovered. In some instances, it can be used for production in the next well. However, in many cases, it must be treated and isolated in lagoons, or pumped down underground for permanent disposal. This latter approach has led to seismic activity in states like Oklahoma, where the number of earthquakes has significantly increased as a result of injection of wastewater back underground.[10]

CHEAP GAS PLUS EFFICIENT TURBINES EQUALS CHEAP POWER

This recent flood of large quantities of fracked natural gas into the market has resulted in historically low-priced gas feeding a large fleet of generators offering low-cost electricity into the market. In some cases, coal-fired plants that were capable of doing so switched to lower-cost, lower-emission natural gas.[11] The result is that existing nuclear and coal-fired facilities find fewer hours each day during which they can successfully compete. As a result, the capacity factors of competing coal-fired plants decrease, they run less frequently, corresponding revenues decline, and it becomes increasingly difficult to cover fixed costs.[12] Nuclear facilities, by contrast, cannot operate at lower capacity factors, but many have been losing money for years. In fact, Illinois nuclear plants only operate courtesy of an annual ratepayer subsidy of $235 million.[13] Similar subsidies exist in Connecticut, New Jersey, New York, and Ohio.[14] These growing cost pressures have resulted in an increasing number of older coal and nuclear plants retiring prematurely, before their originally anticipated closure dates.

While some blame the growing quantity of renewable wind and solar plants (with their zero cost of fuel), the assailant that actually thrust the dagger into the heart of coal country was the enormous success of relentlessly efficient drillers in the oil and gas patch. Yet even as gas generation has pushed many coal-fired plants out of the way, the natural gas fleet faces its own challenges. Gas plants are also now beginning to see lower utilization rates as they face growing pressure from the increasing number of wind and solar plants (as well as battery storage). A quick look around the country at the leading markets for renewables helps tell that story.

AS RENEWABLES CLIMB, GAS GENERATION ASSUMES A NEW ROLE

To understand what's happening, it may be helpful to look at the electricity mix in a few key states. Oklahoma, for example, received 40% of its electricity from wind turbines in 2019, while two other states in the windy backbone of the continent, Kansas and Iowa, stood at 35%.[15] Meanwhile, Texas has enough wind (with some solar in the mix as well) to make it the fifth largest global renewable energy economy in the world. In other states, such as California (19%)[16] and Hawaii (12%),[17] solar is becoming a predominant energy source. Those numbers will keep increasing as the pressure to decarbonize the grid continues to mount.

The challenge, though, is that renewable resources are intermittent. They cannot access piles of coal or gas in a pipeline waiting to be burned. Those tens of thousands of wind turbines and millions of solar panels generate carbon-free and low-cost energy when the weather decides they will, which may or not be when the grid needs them.[18] Globally, the estimate is that the percentage of time that wind and solar resources show up when you need it, coinciding more or less with our peak demand periods—is 14% for onshore wind (because wind often generates more energy at night when the wind is stronger but demand is low), 27% for offshore wind (which has steadier daytime winds), and 20%–40% for solar (since the sunshine is often more coincident with peak consumer electricity demand). These numbers vary greatly by region.

In order to integrate renewable assets effectively into our power grids, three critical time-related issues must be addressed. First, both wind and solar assets are characterized by short-term fluctuations in output that must be addressed in order for the grid to function smoothly and provide the appropriate power quality. When individual clouds pass over a solar array, output can fall off rapidly. I once heard an expert refer to that variable, cloud-influenced output as "PV units running like a seismograph."

If solar arrays are spread out, this effect can be mitigated, since the same cloud cannot be in two places at the same time. But if a large solar power plant is affected, it may be necessary to bring other generating resources on line to make up for that temporary shortfall. A similar phenomenon

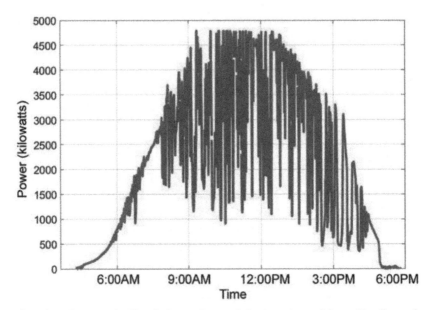

The solar seismograph. *Electric Power Research Institute Journal,* https://eprijournal.com/can-variable-solar-generation-cause-lights-to-flicker/

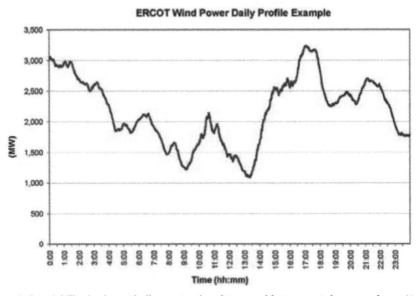

Wind variability is also a challenge. *National Renewable Energy Laboratory,* https://www.nrel.gov/docs/fy11osti/49218.pdf

occurs as a result of the variability of wind, such as is shown in the graph on daily wind output in Texas.

Second, there are difficult-to-predict but inevitable weather events that occur occasionally and can significantly reduce renewable generation. For example, the so-called atmospheric rivers, such as California's Pineapple Express, can bring clouds and precipitation events lasting well over a week and greatly reducing solar output.[19] In the case of wind, a potential front may result in an extended period of calm, with a resulting fall-off in energy generation.

Third, both wind and solar output can be highly seasonal and subject to extended changes in output. With wind, there are typically seasonal patterns with some months much windier than others. With the sun (especially in northern latitudes), the output in the winter is significantly less than in the summer. In either case, other generating or storage resources must take up the slack. Those seasonal differences can be planned for, bringing to bear dispatchable generating plants that are called into action when there is a shortfall of renewable output.

Some of that short-term intermittency can be addressed with the addition of the storage of batteries or other energy storage technologies, although that greatly increases the need to invest new capital. In fact, in just the next few years, developers are planning on adding billions of dollars' worth of batteries. These will mostly be devoted to solar projects to address the issues of both short-term intermittency and the need to shift solar output to meet periods of highest demand. What these batteries won't be able to do is store renewable energy for longer periods, for example from the weekend to a weekday or from season to season. Other technologies are being developed to address that need, but they are not quite there yet in terms of scale and economic competitiveness.

HUNTING THE DUCK

To better understand this level of required flexibility, it may be helpful to examine a day in the life of the grid in California, with its famous solar "Duck Curve." If we follow the supply curve—the net of all solar (and a small amount of wind) generation in the state—we can get a better idea of the increasing flexibility required from the gas plants on the margin (the last ones in and the first ones out) and their challenging operating regimes.

Looking at the graph of a sunny California day in February 2020, we see that even when the state is still cloaked in darkness—from about midnight to 3:00 a.m.—about 2,000 MW of generating resources have to dial back production to meet a decline in demand as most of the world goes to sleep. At about 3:00 a.m., those resources that dialed down now need to increase their production again for a few hours, and others need to join them as demand starts to increase. In the old days, as the world woke up and society began its daily rituals, the amount of fossil-fired generation required to meet increasing daily demand would have arced upward, with a steady climb from 5:00 or 6:00 a.m. right through until late in the day.

Not anymore, though. These days, as soon as the sun comes up, everything changes. First hundreds, and then thousands of megawatts of solar resources—from over a million California rooftops as well as giant solar projects covering tens of acres in California's eastern deserts—begin to quietly pour electrons into the grid as the sun begins to rise. Together, these resources both in front of and behind the customers' utility meters conspire to serve demand that was formerly met by traditional generating plants.

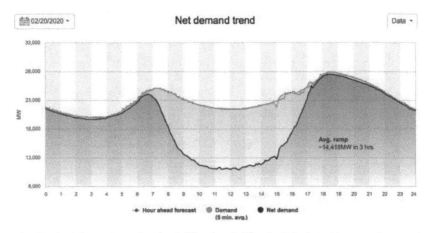

The Duck: A late-winter day in California. *California ISO,* http://www.caiso.com/ TodaysOutlook/Pages/default.aspx

As the sun arcs further into the sky, more direct rays of sunlight stream down onto tens of millions of panels across the state. Faced with this relentless wave of low-cost electrons, the conventional power plants— mostly gas-fired generators—marshal a hasty but organized retreat, forming the duck's increasingly deep belly.[20] One after another of California's 52 gas-fired plants dedicated to meeting peak demand (and that typically operate for less than four hours per day)[21] receive a signal from the California grid operator to ramp down production until, by midday, the amount of required conventional generation may be only half the amount required at midnight. That means that a large portion of the generation fleet must dial back for a period as short as a few hours or as long as 9–10 hours.

On occasion there is so much surplus solar energy that it is sold to other states through an Energy Imbalance Market established in 2014 that "automatically finds low-cost energy to serve real-time consumer demand across the west."[22] In practice, this means that at times California is paying its neighbors to take power because it's cheaper to do that than to shut down some generation facilities.

Then, as the sun begins to move lower in the sky, all those plants must spring back into action, but only for a short period. In this February example, over the course of three hours in the late afternoon and early evening, the required amount of nonrenewable and dispatchable generation more than doubled (thus forming the duck's inclined neck and head). Almost as soon as peak load is met, demand begins to drop and those assets have to start cycling down again, prepared to do it all again the next day. Plants that were originally designed to cycle like that perhaps once every day may now have to do that several times a day. And as even more solar is added to the grid, overall utilization rates may continue to fall.[23] This may especially be the case if large quantities of batteries are introduced into markets such as that of California. In that market, a recent utility study indicated that solar plants with four hours of added storage could ensure that their output would meet the critical demand (defined as "effective load carrying capability") 99.8% of the time.[24]

THE TRADE-OFFS FOR GAS GENERATION

However, until energy storage is installed cost-effectively and at scale, this issue poses a significant hurdle. Gas-fired plants are generally capable of meeting the challenge, but there are a number of important issues and trade-offs associated with this newly adopted behavior. For the gas plant owner or operator, these generally boil down to three critical issues: starts, ramp rates, and minimum turndown. These issues are important to comprehend because they ultimately determine how many renewable resources we can integrate into today's existing systems, and at what cost.

Let's first discuss the concept of "starts"—the process of getting the turbine up and running and burning fuel to generate power. These highly sophisticated gas-fired turbines take in a mixture of air and fuel.[25] The incoming air is pressurized by a compressor, while the fuel from the pipeline is also subjected to pressure, so that a typical turbine sees a fuel pressure of 450 psi.

Depending on the type of turbine utilized, that fuel and air mix is then burned at temperatures ranging from 2,100 degrees to more than 2,800 degrees Fahrenheit—with temperatures at the high end approaching those required to melt steel.[26] The hotter the turbine can run, the more the gases will expand and the more pressure—and thus energy—the turbine creates. This process creates a flow of hot air that departs the turbine at speeds over 150 miles per hour (similar to a category 5 hurricane).[27] Much of the natural gas combustion energy is converted to electrical energy, but exhaust from the turbine still reaches typical temperatures of 1,000–1,100 degrees Fahrenheit.

It's that process of combustion and resulting air flow that gets the rotors spinning inside the turbine at 3,000 or 3,600 revolutions per minute (required in order to match the frequency of the grid). The tip speed of the blades in one of GE's most efficient turbines—the 9HA.02—moves at one and a half times the speed of sound.[28] In the process of getting a turbine up and running, a good deal of stress is created. That stress is much greater if the turbine is moving from a "cold start," where it hasn't been running and the tons of metal in the machine are cool. The operators then have to heat the entire metal mass of the machine relatively slowly in order to bring it up to operating mode—a process that takes close to two hours. By contrast, a "hot start" from a turbine that has been shut down for a period of less than four hours and therefore already warm takes approximately one-half hour.

So, in markets like California with its high percentage of solar resources, or in the middle of the country where wind generation increasingly prevails, many operators find themselves running their plants quite differently than they did just a few years ago. Some have to shut down entirely, while others go into turndown mode more than once a day, where they typically operate at 40% to 50% of full output. Operating at lower loads can lead to lower firing temperatures, which is highly inefficient (peak efficiencies can be cut by as much as 50% relative to full-load conditions). There are also limits at the lower end of the operating range so that plants do not exceed their environmental permit allowances (a rare occurrence, as the machines are set to operate at a level inside their permit window).

But what happens when the renewable "fuel" (sunshine or wind) simply doesn't show up? What happens when you need electricity in California and it has been raining for days, so there is little solar energy to speak of?[29] Or you are stuck in a warm front in Texas and the production of your wind turbines falls off? Transmission lines to other areas not affected by the same weather help to diversify the risk, since one can import electricity from elsewhere, but only to a limited degree.

Today, the only assets providing that compensatory level of reliability are coal and gas plants. Coal-fired plants aren't that much help; because they have high carbon dioxide emissions, they're increasingly uneconomical to run, and—most critically—they are not very flexible. They are difficult to ramp up and down in response to the needs of a grid calling for increasing levels of flexibility.[30]

That leaves us with natural gas. In regions with high penetration of renewable resources, gas plants are increasingly being called on to adopt this new role of backstopping dirt-cheap renewables. These plants were not meant to play that. Regardless, in an increasing number of markets, it's the game they now have to play if they want to be in the market at all.

Operational Trade-Offs, and Cycles Versus Run-Hours

There's another critical consideration as well, having to do with flexibility, since gas turbines fall into two groups: simple or open cycle combustion turbines (a subset of these are known as aeroderivatives, meaning they are derived from jet engines), and combined cycle combustion turbines, which are most often heavy frame engines. Each type of turbine has its own place in the ecosystem.

The smaller aeroderivatives, typically around 50 MW, can ramp up and down quickly as required. They can reach full power from a cold start in 10 minutes. But they operate at far lower thermal efficiencies, hovering in the low 40% range. By contrast, the large, heavy frame CCGTs that are largely favored throughout the utility industry have much higher efficiencies in converting the energy in raw fuel to electricity—well over 60%. However, they are also more challenged by this new environment requiring more flexibility. That's largely because their efficiencies come about as a result of pairing a gas turbine at the front end with a steam turbine at the back end.

Conceptually, the way it works is simple: the pressurized mixture of gas and air enters the 400-ton gas turbine and ignites. The rapid expansion creates a force that spins the rotor within the turbine, generating an initial flow of electricity. This is known as the "topping cycle." The waste heat emanating from the back end is then routed through a heat recovery steam generator (HRSG), which creates high-temperature steam that is subsequently delivered under high pressure to a steam turbine generating additional electricity (the bottoming cycle). The topping cycle typically contributes approximately two-thirds of the combined cycle plant's output, and increases the plant's efficiency by roughly 20 percentage points.

The challenge with the CCGT is that while it is valued for its higher operating efficiencies, it's tougher to cycle the entire machine, in large part because of the large thermal mass of the components in the bottoming cycle. In theory, one can turn down the gas turbine quickly, but there is a lag effect in the steam generator. The further one can turn down the unit and still meet emissions limits, the more valuable these gas turbines can be in balancing the constantly shifting orchestra that is the power grid. GE's newest CCGTs—such as the 9HA.02—have been specifically designed to effectively turn down to as low as 30% load and still meet emissions limits.[31]

Improvements in Turbine Technology

As the grid continues down a path toward an increasing amount of intermittent renewables, the gas technology must continue to evolve, at least until other solutions are found and deployed to balance the needs of the grid with the output of wind and solar resources. Over time, a larger fleet of batteries will help address some of these challenges. In the meantime, engineers at the equipment manufacturers continue their quest to develop more flexible and more efficient turbines that can ramp more quickly and achieve turndowns to lower levels.

A great deal of progress is being made in this area. Historically, it has taken about a decade to increase the efficiency of a CCGT by 1 percent.[32] However, the same technological drive that has led to improvements in wind turbines, solar panels, and batteries is at work here as well. In just a few years, the most efficient gas turbine increased from 61.5% efficiency to 62.2%.[33] Two years later, a GE turbine in Japan achieved 63.08% efficiency,[34] and by 2018, a number of global manufacturers had unveiled turbines with efficiencies exceeding 64%, and 65% or better looks to be within reach.[35] Most of this has to do with heat management, as that is where the improved efficiencies lie. Improved thermal barrier coatings, new materials such as ceramic matrix composites, and highly sophisticated and efficient cooling paths enabled by 3D printing of certain parts—all of these have a role to play in improving efficiencies and operating flexibility.

The other piece of the equation is the growing role of digitalization. GE and other turbine manufacturers are increasingly instrumenting their turbines so that operators have a more precise view into the machines and a better sense of operational trade-offs. Increased operating flexibility creates more stresses, but the addition of sensors and algorithms creates a more accurate view of real-time conditions. In the best case, that allows fleet owners and operators to migrate from a reactive to a more predictive approach.

Digitalization is like an MRI for a power plant, allowing the operators to better understand the actual stresses within the plant so they can make more informed trade-offs with respect to issues such as how hard they can run the plant when prices are high, and how frequently they need to be taken offline for maintenance. A crude analogy for the latter would be changing the oil in your car. Today, we do that according to a schedule based on a calendar or miles driven. But if we had a sensor in the oil that could tell us the precise state of that oil, we'd know whether or not it really needed to be changed. Such approaches can result in a more proactive approach to market conditions and opportunities.

Yet even with an enhanced view, there's no getting away from the simple fact that there is a trade-off between cycles and total run hours. Every CCGT is subject to this same dynamic. The more frequently you start and stop, the more wear and tear you put on the machine. You get roughly 900 cycles, or between 24,000 and 30,000 run-hours, before it's time to take the machine down for an outage that may last as long as a month or more and may involve refurbishing or replacing multiple parts.[36]

It's a new world, and one in which the system is constantly evolving, with continued uncertainty. The scarce commodity is no longer the fuel. Fracked gas is cheap, and the marginal cost of renewables is zero. These days, the value is increasingly in reliability, the ability to balance the system, and the ability to flex and respond rapidly to constant change. From being a system with one entity in charge, dispatching power plants on command, the grid has rapidly evolved into a far more complex system where nobody has total control. And that's today, where renewables account for less than 30% in most grids. It's also before we add in the increasingly robust customer side of the equation, which is about to take off as well.

12

BETTING THE STORE

Corporations, Energy, and Climate Change

The global effort to reduce carbon emissions is never going to occur in a meaningful way if a large number of the world's largest corporations do not become involved. Their activities—whether direct or indirect—are simply responsible for too many emissions for them to remain on the sidelines.[1] Fortunately, many haven't. In recent years, many of these large corporations have stepped up and started searching for ways to reduce their global environmental impact with respect to water and energy use.

As of December 2020, for example, 276 companies across the planet had joined RE100 and made formal commitments to 100% carbon-free energy.[2] Total corporate renewable electricity purchases made up over 10% of all renewable capacity brought online in 2019. Since 2008 those cumulative purchases have exceeded 50,000 MW, equivalent in size to the entire electric generating capacity of a country like Vietnam or Poland. More than two-thirds of the 2019 commitments emanated from U.S. companies.[3]

The growth in this trend has been impressive: In 2014, the first year such information was formally tracked, seven U.S. companies signed eight corporate renewables purchases, totaling 1,200 MW of renewable capacity.[4] By the end of 2019, that number had blossomed to well over 40 companies signing 93 separate deals totaling 9,330 MW.

There's still a very long way to go, since to date there are still a relatively small number of companies that have made the commitment to become carbon neutral. However, the trend is encouraging, especially as it relates to electricity and the purchase of energy from renewable resources.

That's impressive, but what does that mean? Why are these companies spending money on something that's clearly not part of their core business, what are they actually doing, and where is this trend likely to head?

First, let's start with the why: Companies are increasingly finding that their activities are being scrutinized, since they have outsized impacts on the planet. A cynic could say that most companies are only doing this because they have to, because it's about perception. And undoubtedly, perception clearly matters.

However, if you talk to the representatives of some of the companies involved, it quickly becomes apparent that there's something more there. I wanted to better understand this phenomenon, so I lined up virtual meetings with representatives from a couple of key players: Jim Goudreau, head of Climate for pharmaceutical giant Novartis, and Rob Threlkeld, global manager for Sustainable Energy, Supply, and Reliability at General Motors.

NOVARTIS: TAKING THE BROAD VIEW

Goudreau is responsible for all things climate-related for Novartis, which has facilities spread across the globe in 100 different countries on every continent but Antarctica. A former U.S. Navy supply officer for 26 years, followed by stints as acting Deputy Assistant Secretary for Energy for the Navy and director of Policy and Partnerships, two things immediately stand out about Goudreau: one, he understands energy and logistics to the nth degree; and two, he's no bleeding-heart liberal. Pragmatism is the order of the day.

In the Navy, part of his focus was on strategic diversification of the energy supply chain, both in the short term and further into the future. In our discussion, Goudreau commented that since the Navy is tasked with taking a long-term strategic view, the energy conversation there quickly "morphed into a discussion on the impact of climate and regions where the climate was a threat multiplier that in certain cases could lead to physical conflict. The nuance is, climate change doesn't cause physical conflict." However, he emphasized, climate change can create a tipping point, especially in countries with limited resources and weak governance, and "if the system doesn't manage it effectively, through either intent or lack of capacity, then those pressures may tilt the world into instability and conflict."

When Goudreau left the Navy, he was focused on issues related to both climate resilience and sustainability, so when Novartis created the Head of Climate position, he jumped at it. "One of the challenges was on not just climate mitigation, but also on climate adaptation." The company had long been working on cutting greenhouse gas emissions through efficiency investments and even creation of an internal shadow price for carbon, but "it hadn't really started to tackle climate risk and the resilience discussion associated with climate adaptation."

Part of Goudreau's job is to identify which of the pharma company's far-flung facilities are most exposed to the potential impacts of climate change. He cited two locations in California, a research facility in the Bay Area and another down in La Jolla, where each year he needs to determine if they will be affected by the days-long grid outages orchestrated by local utility Pacific Gas & Electric (PG&E) to minimize the potential for wild-fires.[5]

From Goudreau's perspective, "That rate of risk has changed, with more damage to life and property happening as a result. One big game-changer from a perceived risk perspective is the risk of preemptive grid shutdowns. PG&E will continue to do that to minimize risk, and I won't be surprised to see this occurring in other utilities up and down the West Coast and spreading into other areas like the Southeast, especially as we see droughts and events overcoming the capacity of the grid to respond to temperatures and the loads on them. Then that spark dropping to the ground is all you need. This is a new reality that is emerging. That risk triggers economic disruption."

So how does he approach that issue? "Planning," Goudreau commented, "becomes a mixture of approaches. You have to understand what your critical functions are as a company, then you understand where you need to have redundant capacity within your networks. If you don't, then clearly you have to take that level of risk at a facility more seriously."

However, when one begins to take interdependencies into account and widen the lens, addressing the challenge becomes a good deal more complicated. Goudreau cited the example of flooding: "If I invest in equipment that keeps my building dry and the lights on at a certain location, if I can't get my associates to work to perform their jobs, if the transportation network is out, the food distribution network is out, the social network of the community has been torn . . . you can't look at a single issue as a silo. And you can't look at operational capacity without thinking about the industrial supply chains that support that."

Goudreau and his team (of up to nine individuals at any given time) divide their time among helping the company with resilience strategies, efficiency investments, and decarbonization initiatives.

On the clean energy front, one of the company's major efforts to date has been to offset the effect of its U.S. electric power consumption by purchasing 100 MW of electricity from a wind farm in Encerita, Texas, in 2018.[6] Goudreau noted that following this success, the company then began negotiating with developers for a vehicle that would accomplish the same thing for their electricity use in Europe. That purchase will involve multiple European sites, with two to four different physical locations buying both solar and wind energy. His team is also actively pursuing development of on-site solar projects at seven locations.

In the three-and-a-half years Goudreau has been with Novartis, he's seen the amount of time devoted to climate resilience and risk increase. He commented that the Task Force for Climate Related Financial Disclosure "has allowed us to broaden the conversation. There's a greater focus with more people asking about it. Increased reporting requirements are also leading to more robust discussions internally.

"As a company," Goudreau emphasized, "our core mission is to care for and cure patients. That's what we do. The climate crisis is driving a global health crisis. Our duty of care involves a response to the potential and understanding that means we have responsibility to act on climate change as well."

Looking to the future, his goal is to help further internalize that message, so that it becomes ingrained in the corporate culture. "Ideally, he said, "I end up working myself out of the job, and thinking about climate issues and mitigation are part of everybody's informed daily decision-making processes. That's a journey of development we are still on."

GM: DRIVING CHANGE

For his part, Threlkeld has been with GM working on renewables for the past two decades. The company developed its first solar project in 2005. The goal back then was to see if the company could leverage its scale to move the needle in the renewables space, and Threlkeld noted that at the time, the 1 MW rooftop project on its Rancho Cucamonga, California, facility was the largest in the country. That project helped drive scale, generated modest savings, and predated the establishment of corporate renewable goals. Another 1 MW rooftop project in California became operational in 2007. And then the team took it to the next level, with the largest rooftop project to date at that time: a nearly 12 MW solar addition to a GM car assembly plant in Zaragosa, Spain.[7]

Threlkeld marveled at the speed at which the solar costs have plummeted. "In my career, when we did the first renewable project, the price per watt was almost $10, and here we are 15 years later and you are under $1. This gives you the perspective of how markets and costs continue to go down and a lot of that is the first companies back in the 2000s looking to scale this up and drive scale and contract mechanisms."

Threlkeld commented that on-site projects clearly have their place, but they are often limited by the available space, whether it be on the rooftop or adjoining property.[8] So the next big "game-changer" innovation was the purchase power agreement, or PPA, that allowed customers to cost-effectively access renewables off-site. With PPAs, companies could buy the output of large wind or solar projects located virtually anywhere. "There's where the scale hits." he asserted. "It's one thing to do 1 MW or 10 MW on a rooftop. It's another to do a 100-MW deal."

By 2016, GM had already surpassed its original goal of 125 MW, and the question then became, "what is our next goal going to look like?" Threlkeld helped convince GM's leadership to join dozens of other large U.S. companies in RE100, setting a goal of 100% carbon-free electricity supply by 2030 in the United States and globally by 2040. In this context, the PPAs and a similar product—the utility green tariff, through which utilities offered renewable electricity directly to their own customers—became vital tools.

At the same time, there was a concern that everything was focused on renewables, and Threlkeld recalled that "we asked, 'Is that a good position to be as a company?' That's how we came across the four-pillar approach." These include 1) energy efficiency (still the most cost-effective and cleanest way to "green" the portfolio since a kilowatt-hour not consumed as a result of efficiencies is the cleanest one of all); 2) purchasing renewables; 3) moving to an all-electric future with the use of energy storage (fuel cells, batteries, or EVs themselves); and 4) using the company's sheer scale and resources to influence policy that will accelerate adoption of renewable energy.

The pillars are intended to help the company navigate through a future marked by constant evolution, uncertainty, and arrival of new technologies. "We don't know the full journey that's going to take us there," Threlkeld commented, "but we've got an outline of a plan to get us there."

In the sheer "gee whiz!" astonishment surrounding renewables, energy efficiency is often overlooked. However, the reality is that there are many inefficiencies built into the energy consumption landscape. To address that challenge, GM has been making continuous investments in everything from efficient LED lighting to more efficient industrial processes. A key element, Threlkeld noted, has been the use of "all the data that we've got" to better automate and manage energy use along the entire production line from the welding to the paint jobs. One key innovation has been an energy system called Energy OnStar, which offers a supervisory ability to see what all assembly plants are doing from an energy consumption standpoint, and creates the capability to manage in a much faster timeframe. These investments have allowed GM to save over $100 million annually.[9]

On the renewables side, GM has engaged in power purchases or green tariffs across North America. The latest effort was an April 2020, 350 MW agreement with local utility DTE to provide solar energy to its locations in southeastern Michigan, following on the heels of a 150 MW wind agreement in 2019. Together, these commitments will deliver 100% renewable electricity to facilities including GM's Renaissance Center global headquarters in Detroit, its Global Technical Center in Warren, and two local assembly plants—Orion and Detroit-Hamtramck—by 2023.[10] Threlkeld calculated that to date, the company has sourced just under 1,000 MW of a total goal of approximately 3,000 MW.

HOW DOES A POWER PURCHASE AGREEMENT WORK?

When Novartis, GM, or any number of companies indicate that they have committed to a renewables purchase, what does that actually mean? It turns out, it can mean a lot of things. And this explanation is somewhat complicated because it involves a sort of "renewables accounting."

Renewable Energy Credits

The first thing to understand is the concept of "green-ness," which is embodied by something called a renewable energy credit, or REC. A REC is the green attribute from a renewables project, equal to 1 MWh. Let's use a simple example. If I put solar panels on the rooftop of my home, a REC is created for every megawatt-hour they generate (of course, somebody has to track and record that information, and there are specific systems for doing just that). I may decide to keep my RECs, or I may decide to sell them to somebody who wants them. Who might that be? It could be to a utility in a state that requires it to sell a specified percentage of carbon-free electricity (for example, the utility might have a 20% Renewable Portfolio Standard).[11] They could do that by buying green attributes from renewables projects like mine.

If I sell those RECs from my solar panels, somebody else gets credit for my "green-ness." I have—in essence, and put indelicately—"sold my virtue" to some other entity.[12] So the panels on my roof are now just producing generic electricity—nothing I should be bragging to my friends about. I suppose one could call that the industry's dirty little secret, but in truth this concept of RECs has served as an enormous stimulant to the growth of renewables.

Now let's scale that concept up to the commercial and industrial sector. In the early days of the effort to achieve some type of carbon neutrality, many companies bought undifferentiated, "unbundled" RECs. That is, those RECs came from a renewable resource, and they were certified, but one could not attribute the RECs to any single specific resource.[13] So while the purchase of the RECs theoretically offset one's energy consumption with a green resource, it wasn't clear to anybody that this purchase created any real value—it didn't drive new activity or change on the margin.

At some point, companies wanted to be able to point to specific projects that their commitments were responsible for, and to be able to say, "If we hadn't done that, this project would not exist." Thus was born the critically important concept of "additionality."

If a corporation wants to claim additionality and be able to point to a specific project, it can enter into a power purchase agreement, or PPA. This is an agreement, typically spanning 10–20 years, in which the buyer signs a contract with the new project developer. In this approach, the buyer (or "off-taker") agrees to take a specific amount of energy produced by the facility at a fixed price. All of the RECs specified in the contract also accrue to the off-taker, and are created as the associated megawatt-hours of energy are generated. This firm commitment from a credit-worthy entity in turn allows the developer to go to the bank and secure financing—often in the tens or hundreds of millions of dollars, depending on the scope of the project (in fact, the largest project may actually exceed $1 billion).[14]

Physical PPAs

Let's make this somewhat more complicated, because it is. There are two types of PPAs—physical and virtual. With a physical PPA, the renewable resource is located in the same power pool. The off-taker actually takes title to the physical energy at a specified point on the local power grid, and that energy and the electrons can theoretically be delivered to a specific facility (in reality, they are dumped into the grid, like every other generation asset—just like water into a bathtub). Of course, there will always be a mismatch between the time the energy is produced by the renewable asset and when it is consumed by the facility, but the key element is to match the volumes and the associated total of RECs. Physical PPAs are limited to areas of the country where competitive retail power markets exist, in which buyers can actually purchase power from a party other than their utility (which means the entire southeastern, southwestern, and northwestern parts of the United States are generally off-limits, with Texas being one vibrant and notable exception).

Virtual PPAs

With a virtual PPA (VPPA), there is a financial mechanism called a "fixed for floating swap" or "contract for differences" (this is where it gets more complicated) that is settled financially at the close of each month.

Here's how that works: suppose Corporation A signs a PPA with a developer for a fixed rate of $25 for every megawatt-hour generated by the project. What actually transpires is that the wind or solar project sells into the wholesale market at the current spot market (real-time) clearing price, based on the supply and demand conditions during that particular period. If the average spot market energy price over the month (or other designated settlement period) for the project's output is $30/MWh and the developer sells into the market at that price, it owes the off-taker $5 for every megawatt-hour generated (so that Corporation A's effective price is $25/MWh). But what if the average settlement market clearing price is only $23/MWh? In that case, the buyer owes the seller $2 for each MWh under the $25 contract price. This contract-for-differences approach ensures that whatever the spot market clearing price is, the actual value of the output to both buyer and seller is locked in at the contracted price of $25/MWh.

In that sense, the buyer gets a firm hedge against a potentially highly volatile wholesale energy market. However, the critical challenge is to forecast (a better phrase for what is essentially an educated guess) what the wholesale spot market prices will be over the decade-plus duration of the contract, during the most robust period of technological change that electricity markets have ever witnessed.[15]

Such contracts have absolutely nothing to do with physical delivery of electrons, which implies two things: 1) facilities pay their normal electricity bills without respect to the VPPA arrangement; and 2) the facilities whose electricity use is being offset with renewable energy can be located anywhere.[16] For example, in November 2019, McDonalds Corporation announced two VPPAs—a 220 MW agreement to buy energy from Aviator Wind, the country's largest single-phase, single-site 525 MW wind project in Coke County, Texas, and a 160 MW, undisclosed solar project in Texas. Those green attributes will offset electricity consumption from the corporate headquarters and 14,000 restaurants spread out across the entire United States.[17]

There are a number of risks associated with VPPA strategies (in fact, LevelTen Energy, one of the larger brokers in this space, identifies a number of specific risks).[18] While buyers are able to lock in a fixed price, they may find that they committed too early—especially in today's world where solar panels improve in efficiencies by roughly half a percent per year,[19] and the cost of wind and solar have declined by 70% and 89% over the past decade, respectively.[20] That dynamic creates the risk of getting into the market too early, paying more than necessary for the fixed price.

Mechanics of a Virtual Power Purchase Agreement

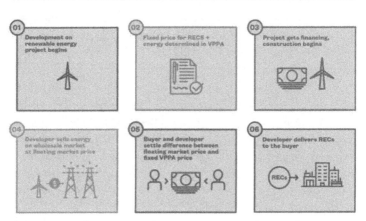

How the virtual power purchase agreement works. *LevelTen Energy,* https://leveltenenergy.com/blog/energy-procurement/virtual-power-purchase-agreements/

Negative Covariance

There's also the nasty little additional risk of what is referred to as "negative covariance," which is a simple way of saying that sometimes, there's too much of a good thing. In California, when there is too much solar energy being produced relative to demand, prices can go negative. The same is true with wind energy in Texas. When that happens, those contract-for-difference checks can be quite substantial. In fact, what happens in Texas matters a great deal to many off-takers because today, an estimated 39% of corporate wind PPAs are associated with wind projects in

Texas.[21] In a grid with just over 80,000 MW of installed generating capacity,[22] 28,336 MW of capacity were wind-generated as of early June 2020,[23] with most of those turbines clustered in the northwestern part of the state.

A perusal of the real-time prices as of 10:45 a.m. on June 5, 2020—as I literally was writing these words—showed that prices in northwest Texas were negative, and as low as $–5.32. That meant an oversupply of wind energy could not be moved by transmission resources out of the region to the areas of demand further south (where prices were higher). As a result, wind turbine operators *were actually paying somebody to take their electricity*. So back to our contract-for-differences example: if the price were to remain at this level for the hour, an off-taker with a $25/MWh fixed price would be writing a check for $30.32 for each MWh specified in the contract and delivered into the system during that hour by the specific wind resources covered under that contract.

In Texas, that dynamic will likely be exacerbated (until energy storage is added to the mix), as there is a significant queue of projects waiting to be constructed and connected to the grid. ERCOT planning documents from April 2020 show anticipated wind resources totaling as much as 37,000 MW by the end of 2021. So, the negative covariance issue potentially gets worse.[24]

Thus, there is this interesting and unfortunate phenomenon that the more corporate entities buy from a place like Texas, attracted by ease of connecting to the grid and a solid wind profile (the wind in the north blows incessantly—especially at night), the softer the prices get, the more the economic value of new and existing resources has the potential to erode, and the bigger contract-for-differences checks the earlier adopters with higher-priced contracts (before wind got as cheap as it is today) must write.

OTHER GREEN PURCHASE OPTIONS

For corporations not located in wholesale markets, but rather in the world of vertically integrated utilities who own everything from the power plant to the distribution poles, there are a growing number of renewables options as well *if* the utility decides to offer them. There are green pricing programs, where the utility will offset your consumption with RECs for a fixed rate. And there are green tariffs, such as the one GM signed with Michigan utility DTE, involving specific projects the utility supports with its own PPA while selling that electricity through to its customers.

In general, all of these have one aspect in common: the output of the renewables facility does not match up well with the hourly consumption profile of the end-use customer. In other words, that customer may be responsible for the additionality that allows its project to get built and pour green electrons into the grid. However, at certain times, its asset is over-producing relative to consumption, and at other times, the actual power consumed is greater than the renewables output. At that time, the facility is actually consuming some amount of power from the grid generated by more carbon-intensive resources such as coal or natural gas.

A NEW FRONTIER: CHANGING
CONSUMPTION TO MATCH RENEWABLE SUPPLY

In 2016, Google—by far the largest corporate buyer of renewable energy—released a white paper entitled "Achieving Our 100% Renewable Energy Purchasing Goal and Going Beyond."[25] In it, Google committed to "work to achieve the much more challenging long-term goal of powering our operations on a region-specific, 24/7 basis with clean, zero-carbon energy." In other words, it ultimately intends to match its energy consumption hour-for-hour with clean electricity resources.

The company hasn't arrived there yet, but what it has done is take advantage of what it does best—managing data centers and computing capability, while manipulating data itself—to change the half of the equation it can best control: its own electricity consumption.

In the Spring of 2020, Google indicated it is developing a new strategy to make data center energy use cleaner by creating a platform it calls a "carbon-intelligent computing platform" that *changes the timing of when* it performs noncritical data processing work.[26] Google has already deployed this strategy successfully in a large data center, shifting noncritical compute assignments to periods of the day when there is more renewable energy on the grid.

How is this accomplished? On a daily basis, Google's carbon-intelligent platform develops and compares two separate forecasts for the coming 24 hours. The first view predicts the expected order in which the grid will dispatch various generating resources—and evaluates the hourly average carbon intensity of each associated generating asset to come up with an hourly carbon intensity profile. Google then creates a second internal forecast that predicts the hourly consumption necessary to perform specific processing tasks for each hour. It also knows which computing load is elastic and can be shifted to another timeframe (helping autonomous vehicles navigate along crowded highways is a task that cannot be shifted, but organizing cat videos is perhaps not quite so time sensitive). Both forecasts are combined to create an optimized hourly computing schedule best aligned with the relative carbon intensity of the grid at any given time. Early results suggest that this carbon-focused load shifting strategy is effective.

Google states that the next, and far more complex, endeavor is to physically shift the compute load between various data centers, based on which grid is less carbon-intensive, and indicates that it will share the results of this effort in research publications.[27]

MICROSOFT—CARBON NEUTRAL IS NOT ENOUGH

Microsoft is approaching the issue of carbon in an aggressive and novel fashion, recently floating the concept of "negative carbon." In January 2020, the company announced its plan to be carbon negative by 2030, with a 2050 goal of removing more carbon than it has emitted since its founding.[28]

In July 2020 Microsoft announced a series of additional steps and commitments, including a new coalition—Transform to Net Zero—with seven other global companies, to help other companies in their respective transitions to net zero.

Microsoft now plans to source 100% of its electricity for data centers from renewables by the mid-2020s, one of the most proactive targets to date, and phase out diesel fuel for backup power by 2030. It is also extending an internal carbon tax across all of its operations, and is requiring suppliers to do so as well.

The company also created a $1 billion Climate Innovation fund, is putting the first $50 million to work to stimulate innovation of new clean technologies, and is committed to develop 500 MW of renewable energy, including investments in communities affected by environmental challenges. Finally, Microsoft announced a path-breaking Request for Proposal to source 1 million metric tons of "carbon removal from a range of nature- and technology-based solutions that are net negative and verified to a high degree of scientific integrity."[29]

For decades, from an environmental perspective, corporations were a source of many of the world's major ills. Many still are. And some are, at least for now, simply engaged in a process of "greenwashing" rather than engaging in steps toward meaningful progress. Today, though, the best corporate leaders "get it."

They understand that a planet that does not guarantee a hospitable climate is bad for their own companies' future survival, and they are beginning to take action accordingly. Some are moving more tentatively than others, but the leaders create the vanguard that others may soon begin to follow in our urgent quest to create a truly sustainable global energy economy.

13

LOOK MA, NO GAS!

I migrated into the world of electromobility both unexpectedly and rapidly. On the afternoon of October 16, 2019, the local weather report had warned of sustained winds from a quickly developing storm that would arrive sometime after dark. When I went to bed that night in my small seaside town south of Boston, the breeze was not that significant. However, by 3:00 a.m. the neighbor's anemometer was pegging out at 102 miles per hour as a "bomb cyclone" tore along the New England coast, toppling a neighbor's massive twin-trunked maple tree onto my 2007 Prius.

Later that week, on a flight to a New Orleans energy conference, I saw an ad on the seat-back television for a Hyundai Ioniq all-electric vehicle, with the promise of $1,000 down and $79 a month.[1] That seemed both too good to be true—and unlikely to last if it were—so as soon as I was on the ground, I began dialing various Massachusetts Hyundai dealerships. After numerous dealers indicated nothing in stock, I finally located a dealership an hour from my home that was taking delivery of 11 EVs from a sister company that had gone bankrupt.

After some calls back and forth, I provided my credit-card information over the phone and ended up leasing a new car—sight unseen—while standing on a street corner in the French Quarter of New Orleans. No test drive, no choice of color (it's silver), but I had a lease for a brand new EV for only $1,000 down, and $131 a month (since I was neither military nor a fresh college graduate, I had to pay more than the enticing initial $79 offer).[2]

My family has had the car nearly a year, and we are impressed with the technology. It drives very well—much better than our old Prius. The quiet hum when you press the accelerator lets you know this is new technology, and it's very noticeable when accelerating onto the highway, or zipping past someone in the passing lane. The torque and resulting acceleration are a profound improvement over the internal combustion engine (ICE), since the energy is transmitted directly and quickly to the wheels.

HIGH (RANGE) ANXIETY

Our Ioniq only gets 130–150 miles of range on a charge, which is on the low end of many of today's EVs. The popular 2020 Nissan Leaf offers up to 220 miles,[3] while the 2020 Chevy Bolt gets 259 miles.[4] Today's indisputable leader among "moderately priced" EVs—the Tesla Model 3—gets between 250 and 322 miles per charge,[5] and its Model S Long Range now boasts 402 miles.[6] In a few years, as batteries become increasingly cheaper, range will no longer be an issue.

While EV technology is superior in many ways, there are still some shortcomings with most EVs that still need to be worked out. To get that 130 miles of range, we charge the car with our standard 120-volt outdoor outlet and extension cord, but it takes about 14 to 16 hours to get the full range. Further, with today's lithium-ion batteries, charging the last 20% of the battery typically takes half the charging time (in order to protect the battery from overcharging, which could reduce its useful life).[7] So if I were driving for a long distance and required multiple charging sessions, I'd only have about 100 miles of useful range.

In cold weather, that number gets further diminished because the cabin heat comes from the battery. Unlike an ICE in which the interior is warmed up by waste heat from the engine (that ought to tell you something about how inefficient an ICE is), EVs draw the heating capability straight from the battery. So, jumping into the car on a 15-degree day and turning the heat up to 70 degrees with the fan on high will drop your effective range by a full 25%. I have experimented with this on many occasions, especially when taking the car on a 60-mile trip to the suburbs around Boston. If concerned about the remaining miles and willing to sacrifice comfort, I can get the range I need to get home by turning the heat down—not a trade-off many consumers are willing to make. Tesla just solved this problem in its Model Y by incorporating a heat pump that efficiently heats or cools the cabin with technology that uses far less energy, but that's currently the exception, not the norm. The point is, the EV technology has great promise but it's still in its infancy.

I have heard charging company representatives argue that range anxiety is an unnecessary constraint holding back the industry. After all, the average American drives about 37 miles a day, a distance that is easily addressed by batteries. And on average, only about 15% of the miles traveled by the average American car are associated with trips over 100 miles.[8] That statistic does not comfort me: it means a big chunk of long trips will involve sitting at a convenience station or some other location like a shopping mall, waiting for electricity to be pumped into the car. If your EV has 250 or 300 miles of range, perhaps that's less of a problem. But with my 130-mile maximum range, I have experienced my fair share of anxiety. And my charging experience hasn't been seamless, either.

One pre-COVID February morning, while doing research for the offshore wind section of this book, I drove first to the Wind Technology Testing Center just north of Boston, and then over to Tufts University, a few miles to the northwest. There, I tried to plug into a public charging network for the first time, and called up the company (ChargePoint) that owned the charging station offering juice at no cost.

When I got back to my car after my meeting, the range hadn't changed, so I assumed I had done something wrong (I never found out what). I drove home (with the heat on low) and arrived with 24 miles of remaining range, and plugged in the car on arrival. However, the next morning the indicator said I still had 24 miles of range. So, I unplugged and plugged in again. No luck.

I had read about Tesla's ability to remotely access its vehicles, and I even had a Hyundai app on my phone, so I naively assumed this would be an easy fix. Before I called the car company, I searched the manual fruitlessly for errors codes and explanations. I also tried online chat rooms, without success.

So I called customer service, and was promptly connected to a representative who requested that I go online with him to jointly peruse the manual. I quickly discovered to my surprise there were no Hyundai customer care staff specifically trained to support the EV platform even though the Ioniq isn't its only EV model. After 15 minutes without success, he indicated that I should call the dealership.

Hours later, a tow truck carted the car away (at no cost to me) to be evaluated at the dealership. Two days later, I had my answer: somehow the time-of-use setting had been activated, which meant that the car would not charge during specific hours (I never found out which ones). The setting is meant to help customers in areas such as California, with its time-differentiated pricing, to keep from charging their cars during the pricey on-peak hours during which electricity might be two or three times more costly than at other times of the day.

LACK OF SALES AND
CUSTOMER SUPPORT IS NO ANOMALY TODAY

Unfortunately, that lack of knowledge and support showed by my customer service representative is not uncommon in these early days of EV adoption.[9] With the exception of a few car companies—most notably Tesla, with its own dealerships[10]—until recently few automakers have been committed to selling or supporting these vehicles outside of California, where about half of the country's EVs are domiciled.[11]

Few dealerships offer electric vehicles, and apparently even fewer sales and support staff can effectively represent their advantages and features.[12] In fact, a November 2019 study by the Sierra Club found that of the over 900 U.S. dealerships surveyed, only 26% offered an EV for sale, and among those, 44% had no more than two vehicles available on their lot. Perhaps worse, in 10% of the cases where surveyors wanted to test drive an EV, the cars were insufficiently charged and could not be driven.[13] In addition, only 0.3% of manufacturer advertising budgets addressed EVs in 2019.[14]

Apparently, part of the reason for this neglect is that dealer profit margins on EVs are slim compared with their ICE alternatives. In addition, sales staff are (justifiably) unwilling to take the time to learn about a technology that is marginal in terms of total units sold (about 1% of all vehicles in the United States). Today, it's just not worth their time relative to the commissions they could be earning in selling conventional internal combustion cars (that will presumably change as volumes rise).

Finally, because EVs have so few moving parts, there are fewer things that can break or that require maintenance. The drivetrain in an EV contains approximately 20 moving parts compared with that of an ICE that typically has many hundreds.[15] There's also no need for oil or air filter changes and regular service checkups. Car dealerships themselves may not have recalibrated around this fact.[16] As a result, the dealership service departments do not benefit from sales of EVs in the same way they may from ICE vehicles.

A SUPERIOR TECHNOLOGY IN MANY WAYS

And yet, it is probably inevitable that the EV will be the vehicle of the near future, simply because it is a superior technology, with better performance, fewer moving parts and, ultimately, less expensive to own on a total life-cycle cost basis than its fossil-fueled competitor. Furthermore, the handful of models available today is expected to expand rapidly, to as many as 500 EV models available worldwide by 2025.[17]

To understand why EVs will win the contest with ICEs, let's start with the performance of the technology itself. EVs are more efficient. According to the Department of Energy, they convert over 77% of the electric energy from the grid to the wheels, compared with about 12%–30% for conventional gasoline-powered cars (most of the losses in the ICE are in the form of waste heat).[18]

EVs also accelerate more quickly and handle better. Some readers may have seen the Tesla video on YouTube in which drivers put a Tesla into "ludicrous mode," which enables the car to accelerate from 0 to 60 mph in 2.9 seconds.[19] As far as acceleration, Tesla and other high-performance EVs such as the Porsche Taycan can outperform most of their conventional competitors.[20] That's because an EV motor generates instant torque, while ICEs need to burn fuel to turn a crankshaft. That simple engineering fact allowed Chevrolet to make the remarkable claim that its somewhat pedestrian Spark had more torque than a Ferrari 458 Italia.[21]

The second advantage enjoyed by EVs is that they generally have a very low center of gravity—a result of the "skateboard" battery configuration that sits in the chassis underneath the vehicle. A Tesla Model 3, for example, carries 1,050 pounds in its battery pack.[22] With all that mass located close to the ground, the car is better able to hug the road and maneuver more adroitly through curves.

In addition, the standard "regenerative" energy recovery braking system (similar to that found in the Toyota Prius well over a decade ago) converts kinetic energy to stored electricity in the battery every time the vehicle slows down.[23] These EV braking systems can recapture about 70% of the energy that would otherwise have been lost with friction-based braking systems.[24] Depending on whether the vehicle is traveling on flat or hilly terrain (going downhill would otherwise require a lot of braking and lost energy that can now be recaptured) or stop-and-go city driving, the total amount recaptured can be as much as 15%–30%.[25]

The final—and perhaps most compelling reason why electrons will eventually hammer the hydrocarbon on the roadways of the future is that battery costs are falling dramatically. In fact, battery pack costs have fallen 87% over the past decade.[26] Even though the cost of some raw materials has risen, the battery chemistries themselves are improving, and manufacturing costs continue to decline.

Although battery cost is the biggest factor in EV prices, generally that hasn't yet translated into automakers delivering cars at lower prices. Rather, it has resulted in cars with bigger batteries and the longer range necessary to satisfy most drivers. The Nissan Leaf—the single largest-selling EV to date—is a good example of that transition, although it has seen a bit of both trends. The 2012 model only got 73 miles, with a 24 kWh battery.[27] By contrast, the 2020 base model costs $4,600 less, but now offers a 40 kWh battery and 149 miles of range.[28]

BUT YOU STILL HAVE TO CHARGE THE CAR

Today, one of the major knocks on EVs is that they take a long time to charge. Drivers are spoiled by the gas station experience, where it takes two or three minutes to fill up a gas tank with 300 to 400 miles of range. EVs don't work like that, at least not yet, and it takes far longer to accumulate the charge in the battery. Bigger battery packs with more kilowatt-hours offer more driving range (on average, about four miles of distance per kilowatt-hour), but the bigger the car's battery, the longer it takes to charge and the more expensive it is.[29]

If you are an average American driver, you probably only drive 30 or 40 miles per day, and your car sits inactive 95% of the time.[30] That activity doesn't require a lot of electricity, and you can easily charge from home. In fact, most U.S. EV drivers do over 80% of their charging at home, and there are currently two ways to achieve that: Level 1 and Level 2 charging.

Level 1 charging uses a standard 120-volt AC plug and generally adds five miles of range per hour of charging—so it won't help you if you are in a hurry.[31] Level 2, by contrast, provides the electricity through a 240-volt plug, the same service used in homes for clothes dryers and electric stoves. Level 2 will provide from 12 to 25 miles of range per hour,[32] but home-owners need to have a dedicated electrical circuit and a Level 2 charger (costing roughly $175 to $700), and a licensed electrician is usually required to install an upgraded circuit.[33] Level 2 chargers are the ones you typically see outside grocery stores, restaurants, movie theaters, or in airport and commuter-rail parking lots.

THE LAND GRAB FOR CHARGING LOCATIONS

Levels 1 and 2 are fine if you are using the car for a typical commute, or running about town. But what if you want to take a road trip? That's where

Level 3 DC high-speed chargers come into play. These can add 100 (or more) miles of range in an hour, but your car also needs to have a special DC fast-charging port.

Tesla was the first company to get out ahead of this, building a dedicated network of fast chargers along many of the main highways across the country. As of April 2020, the company boasted 1,870 supercharger stations across the globe, with over 16,000 individual fast chargers[34] delivering as much as 250 kW of charge.[35]

Tesla's main competitor, Electrify America, recently announced completion of a cross-country charging network from Los Angeles to Washington, DC, with chargers up to 350 kW spaced not more than 70 miles apart. A second charging necklace is being strung from San Diego to Jacksonville, Florida. Electrify America's strategy is to build what it describes as a "future proof network" of these 350 kW chargers capable of delivering 20 miles of range per minute.[36] There are several other major players as well, each building out their networks and snapping up squares on the board.

The "future proof" language adopted by Electrify America aptly reflects the current state of charge. There is no vehicle on the planet today that can even accept 350 kW of charge (for comparison's sake, that's about the same amount of energy used by a large supermarket during its period of highest demand). However, Porsche claims its high-end Taycan can accept a charge rate of 270 kW per hour. While that rate is impressive, it's still not quite the "gas station experience." *Car and Driver* indicates that the equivalent of pumping 15 gallons of gasoline in five minutes would equate to a charge rate of slightly above 6,000 kW.[37]

Some analysts expect we will get much closer to that possibility in the next couple of decades, with one of the national laboratories suggesting we may need to plan for 2,000 MW charging stations within the next decade.[38]

Meanwhile, Chinese EV company NIO raises the question: why bother charging your battery at all, if we can swap it out for you in less than three minutes? The company has built its own independent swapping network in China, and claims to have completed its 500,000th swap as of June 2020.[39] That approach may in fact have been a better and more convenient approach for customers, but for it to be widespread, all car companies would have had to standardize on some basic issues related to size and configuration. And in today's highly competitive world where the battery combined with the computer is the secret sauce for every automaker, that level of standardization would imply a loss of control that is unlikely to happen in the United States.

THE DC FAST-CHARGING TOWER OF BABEL

Adding to that challenge is the highly inconvenient issue quaintly called "interoperability." That is the fact that the various charging networks do not play nice with each other. In order to access them, you often need to have an account with that operator, and some do not accept debit or credit cards, so you need their app or card to charge. This is further complicated by the various pricing schemes offered, with some systems based on the time you spend connected, while others bill based on the actual kilowatt-hours you use, and still others charge simply based on the session. The price per mile can thus be quite variable and difficult to compare (especially when related to the dollars-per-gallon metric, where we have a pretty good sense of our car's miles-per-gallon efficiency and the cost of gas is visible). This is based in part on regional costs and tariff structures that govern the price of the electricity the charging companies must purchase.[40]

These issues exist to some degree because of the ad hoc way the system has been built to date and the inconsistent patchwork of federal and state oversight and coordination. Much of the public-facing charging infrastructure has been inconsistently developed, based largely on a series of state-level grants and utility programs. A few companies also created their own networks with various business models, different proprietary software and approaches to service, and a general distinction between subscribers and nonsubscribers. While about a dozen companies offer public charging stations in the United States, a small handful dominates the space.[41]

Eventually, all the essential elements must be compatible: the cars, charging stations, the networks of the companies that own them, and the related software, so that drivers can access a widespread and reliable network.[42] There will have to be more coordination between the various charging networks. Today, though, most companies operate their own network, with minimal communications or coordination between them.[43] A few have taken the first steps to change that and create a more broadly accessible network, analogous to banks' ATM networks, but there's a long way to go. The long-term vision is that every EV driver will have access to any operator through a common platform, but we're not there yet.

THE PHYSICAL SET-UP IS ALSO A CONSTRAINT

There's also the simple issue of the plug configuration itself. While this is not an issue for common 120 V home charging and Level 2 AC charging,

DC Standard	Connector	Used By
SAE Combined Charging System (CCS)		GM BMW Ford Mercedes Honda Porsche KIA Audi Hyundai VW
CHAdeMO		Nissan Mitsubishi
Tesla Supercharger		Tesla

Today's DC charging complexity. *"Interoperability of Public Electric Vehicle Charging Infrastructure," Electric Power Research Institute,* https://www.eei.org/issue-sandpolicy/electrictransportation/Documents/Final%20 Joint%20Interoperability%20Paper.pdf

it's still a concern for high-speed DC charging. There are three different charging configurations, with three different types of ports in use on various EV models.[44] It's as if we needed three different types of hoses and nozzles at our gas stations.

The combination of different infrastructures and the various charging networks leads to a situation in which drivers who need a charge from networks they are not affiliated with must currently set up a new account with a vendor prior to charging, through an online sign-up process or a telephone conversation.[45]

Then there's the challenge of finding the stations with the right port configurations. With each proprietary network partitioned off from the others, drivers may have difficulty finding out where charging stations are, and whether they are currently occupied—a critical element, since fast charging to obtain significant driving range can be a lengthy process of a half hour or more.[46]

If we had this same issue in fueling with gasoline, it would mean that somebody with an ExxonMobil or Shell card could not fill up anywhere but at one of those company's stations, might need a special nozzle, and could wait half an hour for the person ahead to fill the tank. That's a lot to think about when one's car is supposed to be a hassle-free way of getting around town.

"FOR LONG TRIPS, IT'S NOT THERE YET"

I decided to see if this was only a theoretical problem, or something to take more seriously. On the Sunday of Memorial Day weekend, our family jumped in our Prius (the Ioniq would have been the better drive, but we didn't want to spend any time charging) and headed from south of Boston to Albany on a quick round trip to pick up a friend. There are currently three charging stations on each side of the 138-mile Massachusetts Turnpike (I-90) from Stockbridge to Boston,[47] located at miles 8.5, 80.2, and 117.6 if one is heading east (with similar locations on the westbound side). These were installed in late 2017 by the Massachusetts Department of Transportation "in order to enable long-distance electric vehicle travel, increase range confidence, and support electric vehicle adoption throughout the Commonwealth," and were originally free of charge.[48]

Since we were already going to be passing all of the stations, I prevailed on my family to let me undertake a bit of highly unscientific citizen science, pulling into each rest area to see if anybody might be charging and their impressions of the experience. The first two stations I checked going westbound had nobody at the 50 kW Level 3 high-speed chargers, which was a bit disappointing.

Heading back eastbound, I came up empty in Lee, Massachusetts, but my luck changed on my fourth attempt as I pulled into the station at Charleton service area, and eased my 2009 Prius past the 24 gas pumps, to the fast charger.[49] There, I noticed a dark blue Nissan Leaf with the front hood slot open for the charging cord and a man standing next to the car, with his EVgo card in hand, talking through a blue hospital face mask on the phone. I donned my own mask, went over to the passenger side where a woman was sitting, and introduced myself as somebody looking to find out more about the charging experience. "You'll have to ask my husband," she replied. "He's the one who takes care of that."

As I waited to chat with him, it became clear that he was having an issue getting the charger to work. The problem was this: the charging station has two chargers, each with a different configuration. The one needed for his Leaf (called CHAdeMO) was not working, and there was no redundancy. The gentleman (I didn't get his name, partly because of the mask) indicated that his Leaf had about 200 miles of range, and he had about 100 miles left.[50] He had been planning on acquiring 70 or 80 miles of range to complete the last part of his trip, but was out of luck.

So now he was considering driving roughly five miles east and then back to the west to access the charging station on the other side of the road, but that would require an additional 10 minutes of travel. His next recourse was to see what charging stations existed on Route 495 northbound. So how did this affect his perception of the EV experience? As it turned out, this was not his first EV rodeo, nor his first Nissan Leaf (his prior one—a 2017—had about half the range). He usually charged at home with a Level 2 charger, and has loved both cars for commuting. This attempt to extend beyond his normal range wasn't going too well, but he told me that in a normal, pre-COVID-19 world, he and his family would have charged for about half an hour, and relaxed. "I might get 100 miles in 30 minutes. And we'd go get some food. Now," he pointed to his mask, "we stay in the car." How would he characterize the experience of owning and driving an EV? "It's OK for commuting," he commented, "but for long trips, it's not there yet. Maybe with a Tesla," he mused. The next day, I pulled up the charging site online and it was listed as out of commission. A few days later, I checked again, and it was working.

Some may read this and say, "See, this whole EV thing doesn't work. They are unreliable, and not worth the effort." However, today's status is not tomorrow's reality, and there is one simple underlying fact: these are nothing but growing pains. This technology is only going to get better and more cost-effective. To take just one example of that pace of change, battery costs fell 50% between 2018 and 2020, so EVs can go longer distances for a lower sticker price.

The latest prediction is that EVs will be cheaper to own than ICEs by about 2025.[51] Batteries are going to become far better and more powerful, and the days of conventional gasoline engines are numbered. In fact, analyst Bloomberg New Energy Finance (BNEF) recently made the astounding claim that "global sales of internal combustion engine, or ICE, cars peaked in 2017 and will continue their long-term decline after a temporary post-crisis recovery."[52] Within a few years, the number of EVs sold will increase significantly, which suggests that soon there will be many more charging stations on our thoroughfares.

Of course, it is undeniable that we won't get the same range for each minute of our fill-up experience. However, chargers will get faster as well. There will indeed be a day in the foreseeable future when those 24 gas pumps are outnumbered by their electric siblings.

THE WAY WE CHARGE OUR
VEHICLES MAY CHANGE AS WELL

Today, whether on the highway, around town, at work, or at home, EV drivers physically plug in their car. But that scenario may ultimately change as well. Soon, we may be able to charge our vehicles without touching a cord or an outlet as inductive wireless charging becomes more common-place. In order to see where this part of the industry may be going, I took a virtual trip to Jeremy McCool's mother's garage in Alamogordo, New Mexico.

McCool is founder and CEO of HEVO, a wireless EV charging startup company.[53] In June 2020, McCool was a COVID-19 nomad, hav-ing moved temporarily from Brooklyn, New York, to New Mexico. In today's world, that meant he hadn't missed a beat, and as we spoke by teleconference, he walked me through his connected systems and what they may soon mean to the EV driver of tomorrow.

McCool has been at this task for almost a decade, and like many im-passioned entrepreneurs, his trip has taken some interesting twists and turns. He started HEVO in 2011 after returning from Baghdad, where he served as an infantry platoon leader, and while getting his master's degree in public administration at Columbia University.

While in Baghdad, McCool's team was assigned the task of getting power back to a sector of the city. Neither he nor anybody in his platoon knew much about electricity, but McCool is not the kind of person who lets that sort of triviality stand in his way. He studied the local power grid and soon determined that his area wasn't going to be served by the central-ized grid anytime soon. He then set out to build a microgrid funded by the local Iraqis, and within 24 hours of the system being up and running, they had "wires running all over the place" to power their various devices.

That experience drove him to focus on improved energy solutions, and HEVO was born in New York while McCool was completing his graduate program at Columbia. For years, the company was quite literally "two guys in a garage." Now, almost a decade after launching, the com-pany offers a full complement of technologies from the inverter on the wall to the wireless charger on the ground, to the wireless receiver in the car, and to the app that links it all together.

During my virtual tour, McCool started by pointing his phone camera at the black pad on the garage floor, which transmits 7.2 kW of energy to his "ancient" black and dinged-up Nissan Leaf, at about an 88%–89% efficiency rate (fairly similar to that of most cord-based charging systems).

HEVO uses an electromagnetic technology, which means it oscillates on a fixed frequency between the transmitter in the pad and the receiver in the car, the same way your electric toothbrush does when it sits on the stand. Inductive charging is not a new technology—Nikola Tesla showed that it could be done in 1893.[54]

The polycarbonate power pad reinforced with an aluminum bezel is slightly under a square meter in size, fire- and water-rated, American Disabilities Act–compliant, and drive-over rated to 1,100 psi (in other words, as strong as concrete), as tested and verified by UL. It has a thick wire conduit that trails across the garage floor to the impact-tested 240 volt inverter box on the wall (which can be lit from inside with LEDs, so that it can be branded by third parties).[55]

The secret sauce, McCool commented, was in the "power station" box that converts AC electricity from the local power grid to the DC that the car battery needs. The box also communicates all of the necessary information to the related IT ecosystem that oversees everything and ensures functionality—which includes the app and the overall cloud-based network, in which 85 different data elements are streamed every five seconds. These include temperature, frequency, and voltage, all of which are time-stamped, recorded, and stored.

In order to receive a wireless charge, the EV must be outfitted with two pieces of hardware: a wireless system nested up in the front of the car under the hood, which connects with the power system, and a receiver in the bottom of the vehicle, which is covered with a plate to protect against damage while driving. McCool lay down on the garage floor during our conversation, and extended his phone camera under the vehicle so I could see the shield, commenting that he has tested that extensively (which explains the dents on the car). He drives frequently through the mountains just outside of town, and noted, laughing, that he had "literally high-sided right on it just last week," so the weight of the car was briefly and largely on the plate underneath.

He climbed into the car and backed it out of the garage while turning on the app so I could see it over Zoom. He showed me a Google-type map highlighting the location of the charging station—something we're all familiar with—and then the tour got interesting. The app began doing its thing once it approached within 25 feet of the power station, displaying visual and audible cues to help McCool close in as he approached the charger (this is also meant to interact with future autonomous driving systems).

On the screen, the pad showed up as a large red dot. At the same time, the sweet spot in the car was depicted as an open circle. When the two

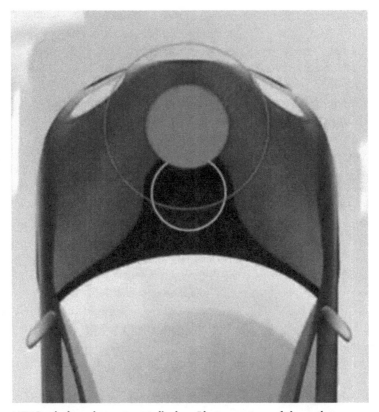

HEVO wireless charge target display. *Photo courtesy of the author*

aligned, presto! Although not perhaps not quite so majestic, it reminded me of the famous scene in Kubrick's *2001: A Space Odyssey* when the Pan Am shuttle docks with the space station, with both systems syncing up as the spacecraft approaches its intended target.[56]

After nearly a decade in the making, the critical technologies are now in place. But it's the business model that always matters most. Here, McCool appears to have figured out the critical ingredients. First, he's outsourced production while developing an approach by which the equipment will be financed by a third party, similar to the solar leasing model, so HEVO will be paid upfront for the equipment. The EV driver/customer will sign up through a typical subscription model (much like customers for solar or telecommunications, for which the device is included in the monthly fee).

McCool's long-term plan is to broaden the company's reach by creating interoperability with other charging, cord-based networks that exist in the market. McCool also plans to roll out a wireless bidirectional charger within 18 months, so that car batteries can actively supply power to the grid when paid to do so.

One estimate suggests a global market of close to $3 billion for wireless EV charging by 2030.[57] Wireless, interoperable, and bidirectional charging (not to mention autonomy)—it's a mouthful, but it may well be the future for at least part of the EV industry. Get used to it.

BUT ARE EVS REALLY CLEANER?

One question frequently asked is whether EVs are really cleaner. What about cars charging up on a coal-fired power grid, and what about the environmental impacts from inputs such as the batteries? To best understand the answer to that question, it's important to tackle this by looking at things on a total life-cycle basis—from the environmental impacts of all the inputs to the actual emissions of the vehicle themselves.

With any analysis of this type, it's worth bearing in mind that assumptions matter greatly. That said, a very recent (2020) and fairly exhaustive analysis evaluating 59 grids around the world found that only in the dirtiest coal-intensive grids, such as Estonia and Poland, would EVs have a negative effect with respect to emissions.[58]

Unsurprisingly, one input that affects the overall life-cycle emissions profile has to do with where the batteries are made. That's the single greatest differentiator in the manufacturing part of the equation, and batteries tend to be energy-intensive because of the mining and refining of the minerals such as lithium and cobalt, as well as the need for extreme heat and sterile manufacturing conditions. Thus, making batteries in a cleaner power grid has a big environmental impact. An analysis of multiple studies concludes that "although the manufacturing of batteries does not outweigh the life-cycle environmental benefits of electric vehicles, these emissions are nonetheless substantial."[59]

ARE EVS SAFER?

When an EV does catch fire, it often makes the news, in part because the fires can be hard to extinguish. However, this must be put into perspective. The National Transportation Safety Agency indicates that "the propensity and severity of fires and explosions from the accidental ignition of flammable electrolytic solvents used in Li–ion battery systems are anticipated to be somewhat comparable to or perhaps slightly less than those for gasoline or diesel vehicular fuels. The overall consequences for Li–ion batteries are expected to be less because of the much smaller amounts of flammable solvent released and burning in a catastrophic failure situation."[60]

That's good news, but the National Fire Protection Agency (NFPA) indicates that there are some issues we need to be concerned about with respect to batteries and fire. A recent article in the NFPA bimonthly journal relates a particularly concerning episode, with implications for energy storage on the power grid as well. On the morning of March 23, 2018, a Tesla Model X slammed head-on into the concrete median divider separating Highway 101 from the off-ramp in Mountain View, California, killing the driver.[61] Within a few minutes of the crash, which was caused by the driver's overreliance on an Autopilot system, the Mountain View fire crew arrived, and quickly recognized that it had a problem.[62] The Model X contained 75 kWh of batteries—enough to power a home for two-and-a-half days—packed into the protected metal case beneath the car. These batteries were organized into a dozen or so modules, containing hundreds of cells in each (the cells are cylindrical, and look somewhat like AA batteries).

The first responders knew they were facing something quite different than a conventional car wreck, since lithium-ion batteries have some unique properties when exposed to enough heat. In particular, when damaged, these batteries can release significant amounts of their own heat, which in turn can propagate to other cells in a phenomenon quaintly called "thermal runaway." That phenomenon can result in off-gassing, fires, arc flashing,[63] and sometimes explosions. Not something the fire crew wanted to deal with on a crowded highway at the end of rush hour.[64]

This unfortunate situation was mitigated by an unusual coincidence: it was only 21 miles from Mountain View to Fremont, where Tesla's headquarters is located. So the Mountain View firefighters summoned a team of Tesla engineers to assist. Tesla's team undertook the arduous job of pulling apart the damaged battery assembly, one cell at a time, and dropping the cells in buckets of water to cool them down. Roughly four hours after Tesla's arrival—with the highway shut down the entire six hours, and firefighters working with hoses to cool the batteries down and keep the fire from reigniting—the team had removed one-fourth of the batteries. The vehicle was then deemed safe enough to remove from the highway and was towed about 20 miles to a San Mateo impound yard, followed by a fire engine. End of story.

Except it wasn't. Within the following 24 hours, the salvage yard had to summon San Mateo's firefighters twice because the vehicle reignited, and they successfully put it out. End of story.

Except it wasn't. Six days following the accident, the car's battery caught on fire yet again, with the problem eventually being solved by Tesla's engineers, who pulled the battery from the chassis and sunk it in a large vat of salt water. Later that same year, another Tesla caught fire, and reignited hours later in the tow yard after the initial blaze had been extinguished.

The *NPFA Journal* stated that "the crash—and dozens of others like it around the world—lay bare the gaps in our understanding of what can occur in such incidents, and how far responders have to go to prepare for the rapid influx of battery technology that experts say is fast approaching." Today, the strategy taught by the NFPA is to pour "copious amounts of water" directly on the area of the battery case, and use a thermal imaging camera to periodically look for signs of heat from the ongoing chemical reaction inside the battery."[65]

Other countries are taking the approach an imaginative step further. Dutch firefighters deploy tow trucks to bring shipping containers to EV accident locations, fill the containers with water, and then use cranes to drop the car into the bath. The NFPA comments that this tactic is becoming more widely adopted on the continent, with ladder trucks being converted into cranes capable of lifting EV wrecks for just this purpose.

CONVERGENCE BETWEEN THE AUTOMOTIVE AND UTILITY INDUSTRY IS INEVITABLE

In a high-demand scenario, by 2050 electric vehicles could increase total electric consumption by as much as 38%.[66] At the heart of this issue is the fact that an EV employs a huge electric battery on wheels. Thus, it becomes important to optimize *when* and *how* these vehicles are engaged with the power grid. Charging at the "wrong" time of the day—coincident with maximum demand on the system, for example—could result in the need to build more power plants, transmission wires, and distribution poles, at a cost to society. It therefore makes sense to charge during the right times of the day (like my Hyundai was reprogrammed to do, without me knowing it). This often results in cutting fuel costs for the driver, while making the grid more efficient as it delivers more electricity over the same fixed infrastructure.

The optimal charging times may vary considerably by region, and those times may change over time as well. For example, in the early 2010s, the concept in California was that vehicles should charge at night during periods of lowest overall electricity demand. That would essentially help flatten the demand curve and make the grid more cost-effective. Just five years later, with the massive influx of solar power, including over a million solarized rooftops, the entire scheme shifted.

The infamous California "duck curve" examined in chapter 11 shows net energy demand sagging during the middle of the day—a result of all the solar-generated electricity pouring into the grid. During those sunny days, the best time to charge is during the height of the day when, in some cases, that solar power would simply be curtailed (wasted), essentially raising the net demand curve and "tummy tucking" the duck. In most other grids, nighttime power is still the less-expensive option and charging is best undertaken during that period. By programming EVs to charge during optimal periods, they can perform a valuable role in increasing the overall utilization (load factor), much like adding more passengers to a flight that would have occurred in any case. But that's only the first small part of an enormous evolution that may dramatically transform the grid as we know it.

VEHICLE-TO-GRID MAY CHANGE EVERYTHING

Already, a small number of EVs are taking the first tentative steps to further interact with the grid, by moving energy not just from grid to vehicle, but in the other direction as well. In 2019 in Japan, Mitsubishi introduced its "Dendo Drive House" concept that integrates the home with the vehicle in "a new energy ecosystem allowing owners to generate, store and share energy automatically between their cars and home."[67] For its part, Nissan just created a program where drivers of its Leaf can pay for parking with the energy from their batteries.[68] In Ontario, software startup company Peak Power supports a fleet of commuters who plug their Nissan Leafs into two office buildings through bidirectional chargers, in a pilot project to deliver juice into the building and offsetting high-priced services from the power grid.[69]

Volkswagen also sees this as a huge new area for potential growth, with its chief strategist citing the ability to sell power back to the grid. By 2025, the company expects to have "350 gigawatt-hours of energy storage at our disposal through our electric car fleet. Between 2025 and 2030 this will grow to 1 terawatt-hour worth of storage. That's more energy than is currently generated by all the hydroelectric power stations in the world. We can guarantee that energy will be used and stored and this will be a new area of business."[70]

Tesla is actively working on this concept as well. It recently developed a platform called Autobidder that utilizes machine learning to allow "independent power producers, utilities and capital partners the ability to autonomously monetize battery assets."[71] In other words, it bids the capabilities of batteries into the power grid to provide various services, without the need for humans to get involved. Today, that capability does not yet extend to the batteries in our cars, in large part because frequent interaction between EV batteries and the grid would deplete batteries whose current chemistries only facilitate 500 to 2,000 cycles.

However, that may be about to change in very dramatic fashion. Tesla recently announced batteries that will offer over 1 million miles of range, and GM is also working on a robust battery platform.[72] Meanwhile, Chinese manufacturing giant Contemporary Amperex Technologies has already announced a 1.2-million-mile battery at only a 10% premium.[73] The average EV battery pack offers about 150,000 miles with today's warrantees, so this is an enormous upgrade in terms of what it offers the driver.

Perhaps even more important, it entirely eliminates the restriction on vehicle-to-grid applications. Today, Nissan's Leaf and Mitsubishi's Outlander are the only EVs that actively support V2G technology.[74] In fact, Nissan just announced in early August 2020 that drivers could pay for parking at its Yokohama Pavilion with the energy from their cars, a demonstration of its Energy Share V2G system.[75] With a supersized battery, in the near future everybody can get into the game.

In my interview with Rob Threlkeld, GM's global manager for Sustainable Energy, Supply, and Reliability, he painted the future that many automakers are considering: "I think a lot of it comes from the power over your cell phone. You'll have connectivity to vehicles and appliances in your home in real time to know your consumption and your rates and what's using what. You will have much more control over how much you pay for the electrons coming to your house. . . . And with a vehicle battery lasting a large number of miles I can say, 'I'm only going to drive ten miles tomorrow, so I can discharge the battery 50% and not worry.'"

Threlkeld predicts that our cars will soon provide multiple value streams to ourselves and to the power grid. EVs and their batteries will be programmed to flex, charging or discharging in order to integrate increasingly large fleets of wind and solar plants. "You'll see much more vehicle-to-grid, whether vehicle to home as power backup, or vehicle to grid, or wherever. Your EV will be part of a broad ecosystem to maximize technology."

In other words, customers will have a greater ability to control how and when we use that electricity. To get a sense as to how this future might evolve, I reached out on a Zoom call to Thomas Ortman, vice president of VoltaBox North America. A big conceptual thinker, Ortman discussed just how significant these changes may become. "Vehicle-to-grid is another income stream, to either the car owner or the manufacturer, and you start to see whole economics change. The truth is, when EVs become less expensive than internal combustion vehicles, you will start to see this tipping point that becomes a landslide and tsunami."

From Ortman's view (and that of numerous observers), most automobile manufacturers are already pretty far behind Tesla because that company got out of the box early, with a laser focus on creating the entire integrated energy ecosystem from solar power (its Solar City acquisition) to in-home storage (the Powerwall) to cars, grid-scale batteries, and the best software in the industry. That has been Musk's vision almost from the beginning.

This coming EV revolution will be bigger than most of us realize, and Ortman believes it has profound implications for our power grids in a number of different ways. First, he said, is the direct impact of EVs on the automotive industry, when Tesla and others "sell more cars and eviscerate the internal combustion business."

Ortman cautions that society has not yet truly come to grips with what this level of disruption will mean, when EVs take over the mobility landscape and affect the supply chain and entire repair and maintenance aspect of the industry. "Think about the businesses that are going to be affected by that. If you make brake shoes, you are never going to sell another brake shoe. If you are in the auto parts supply chain, making the thousands of parts that go into an engine and transmission, you had better be looking for the next thing. We will need a massive commitment to help workers transition from one century to the next century."

Then there's the ability of the new vehicle fleet to potentially provide massive amounts of energy storage to the grid—storage that has already been bought and paid for in order to support the primary mission of mobility. "The ability to build dedicated utility-scale energy storage is one thing," Ortman commented, but you also have "the ability of using all of these cars out there that are not doing anything most of the time . . . the grid could take advantage of all these electric vehicles with smart systems that allow you to charge when helpful and discharge when needed. You'd need artificial intelligence, blockchain, and incomprehensible amounts of data. This is to me a twenty-first-century embodiment of what a grid could be and should be."

Finally, though, one has to think about the sheer volumes of electricity such a transition would require. Recent estimates suggest that the U.S. EV fleet may grow from 1.6 million cars on the road today to somewhere between 10 million and 35 million vehicles by 2030—admittedly a very wide range.[76] Assuming a middle case of, say, 20 million vehicles, mostly passenger vehicles and light trucks (for reference, there were roughly 280 million vehicles registered in the United States at the end of 2019), what might that imply in terms of its impact on the power grid?[77] One recent analysis suggests that it would involve expenditures on the order of $75 billion to $125 billion, covering a wide range of costs including up to $50 billion for new power generation and energy storage, up to $25 billion in upgraded wires and poles to move the necessary power, and as much as $50 billion for the chargers and associated infrastructure.[78]

To put that last number into context, less than $2 billion has been approved for utilities to build out the necessary EV charging infrastructure as of mid-2020.[79] It's no wonder that large oil and gas companies such as Shell and Total are making acquisitions and moving into the space, while the utilities are eyeing the prospects with both exhilaration and apprehension.

As VoltaBox's Ortman framed the issue, there will be multiple challenges to the grid, all taking place at the same time: "Not only do I have to take coal (for economic reasons) and natural gas (for climate reasons) out of the equation, I have to take petroleum out of that macro-energy equation as well, and replace that supply with electricity that will be have to be generated somehow." Petroleum represents 26% of all energy consumed in the United States, "and I'm trying to shove that into the electric power generation component. . . . It's doable, but the numbers are so large that nobody really understands them. Nobody can really understand what 100 million barrels per day consumed means. We can't wrap our heads around that."

It's not like that adoption rate will be equal across the country, either. Today, for example, California has fully 50% of the EVs on the road, with a small number of additional states making up most of the remainder (12 states, with 30% of the country's population have zero-emission vehicles sales mandates). These states will face the greatest near-term challenges and opportunities as we gear up for the decarbonization and enormous transformation of our transportation sector.

Ultimately, if the next car that most of us buy isn't powered by electricity, there's a very good chance the car after that will be.

14

FUTURE IMPERFECT

On December 23, 2015, the sun was just setting at 3:30 p.m. in the Ukraine. A utility control room operator at his desk watched in surprise as the cursor on the screen began moving on its own—an unwelcome ghost in the machine—and clicked on the box to open breakers that would interrupt the power flow at the substation. He watched as a window popped up asking him to confirm the activity, and the cursor—apparently moving on its own—clicked yes. Fruitlessly, he moved his mouse over the cursor in an attempt to retake control, and just then he was logged out of the system. Unable to log back in, as his password had been changed, he could only watch his screen as one breaker after another was opened, cutting power to additional areas of the distribution network as winter's darkness fell upon the landscape.

A LOSS OF SITUATIONAL AWARENESS

That day, cyber-assailants took an estimated 30 substations out of commission at three different utilities, rendering almost a quarter-million inhabitants powerless. The attackers also remotely cut the backup power to two of the three control rooms, literally leaving them in the dark as well.[1] In addition, the hackers launched a denial-of-service attack on the utility customer care centers, with thousands of telephone calls cascading into operators, tying up the lines, and effectively depriving the utilities of any ability to gain a view of what was happening on the ground from customer outage complaints. From the control rooms to the customer care centers, the utilities' situational awareness was completely compromised.

Although the coordinated attack happened in mere moments, the groundwork had been laid for many months, starting with a successful "spear-phishing" effort in May to hack into the individual computers of various employees by using malware joined to various attachments. Once they had breached the initial IT defenses, the hackers set about the work of reconnaissance, moving into what is known in the cyberindustry as "Advanced Persistent Threat" mode. They observed the networks, stole operator credentials, and eventually put the plan into action.

Perhaps the biggest challenge for the hackers was to migrate across the firewall separating the internal networks with Internet access from the operational technology (OT) systems and burrow into the OT networks. Those are the closed systems that oversee and control the network of generation plants, transmission lines, and transformers, and they are referred to as the Supervisory Control and Data Acquisition (SCADA) network.

Even though the outage lasted a relatively short time as operators went out and manually closed all the breakers (no location was without service for more than six hours), the effects were longer-lasting, as firmware on some devices was overwritten so that these were incapable of responding to remote instructions. That had been done so that operators could not remotely reclose the breakers after the initial interruption and any remedies would need to be performed in the field.

This was the first cyber-assault on a utility, and a wake-up call to grid operators and utilities around the globe. It was not, unfortunately, the last. Just a year later, a Ukrainian utility was once again the target. The December 2016 attack was far more limited in a physical sense, affecting a single transformer for about an hour. But it was seen as far more serious for a number of reasons. First, unlike the initial attack, which involved individuals actively engaged in the physical manipulation of cursors and systems, this approach deployed a sophisticated and modular toolkit that allowed it to be both automated and highly scalable. A detailed review of the code also suggested that this episode—which failed to some degree owing to coding errors—involved an intent to destroy physical equipment. The attackers specifically designed their approach to initially disrupt operations and then disable critical equipment, such as protective relays, compromising situational awareness in the field. They also knew from the 2015 event that personnel would likely be in the field in manual restoration mode, so when the breaker was closed and power could once again flow across the system, the protective systems would not be operational. Not only could equipment have been badly damaged, but personnel lives were also at risk.[2]

In today's world, the comforting and familiar distinction many of us have lived with for years—that the battlefield will always be somewhere distant—may no longer be true. In fact, the next field of conflict may be in preparation as you read this, and that field is our own civil society. Our complex power grid and other interdependent systems such as communications, transportation, water, and gas supply (many of which also rely on SCADA systems) may be the next targets, ones through which an adversary could inflict tremendous—even existential—damage. It is important that we recognize this and take steps accordingly.

Fortunately, U.S. utilities have thus far been spared the types of attacks inflicted on the Ukraine, but they are far from immune, and there is already abundant evidence of both our vulnerability and the hostile actors that mean to take advantage of it. A 2020 review from leading cybersecurity firm Dragos characterized the North American utility industry as "a valuable target for adversaries," and noted that the "threat landscape focusing on electric utilities in North America is expansive and increasing, led by numerous intrusions into ICS (industrial control systems) networks for reconnaissance and research purposes and ICS activity groups demonstrating new interest in the electric sector."[3] It identifies seven state-sponsored groups specifically targeting the North American utility sector.[4]

Incidents in recent years are unsettling. In 2013, Iranian hackers intruded into the SCADA system of the small Bowman flood control dam in Rye, New York, where they kept the doors open.[5] In 2017, Russian hackers reportedly breached numerous U.S. electric utilities through their vendor networks, using tools that duped vendors' employees into entering their credentials that allowed hackers a foothold into the solutions providers and eventually into the utilities themselves.[6]

Over the next two years, from 2018 to 2019, 17 smaller U.S. utilities became targets of a Chinese hacker group, victims of a phishing email campaign to infect employees' computers with a remote-access malware program called LookBack.[7] And in March 2019, a new type of denial-of-service attack in the western United States occurred when an attacker continuously rebooted a utility's firewalls meant to protect the system. Operators thus lost visibility into specific areas of the grid every time these devices cycled, over a period of approximately 10 hours, though no physical flows of power were affected.[8]

Clearly, the threats are real, and—based on reports from the experts—they appear to be growing.[9] Part of the challenge is similar to that faced by the defensive unit of a football team that must defend against every potential offensive play, in a game where the offensive playbook is continuously being developed. It's hard to keep the offense from scoring, because the defense doesn't know what will come next. That challenge may grow significantly with the introduction of artificial intelligence (AI) into the arena. AI can help attackers more quickly develop new schemes, and may also help attackers better blend in with their environment on the systems they are surveilling and attacking.

One cyberfirm that utilizes AI in defensive strategies for clients paints a challenging future, in which "a highly effective use of machine learning will be to train malware in optimal decision-making. . . . Supervised machine learning can transfer the skills of the best malware operators directly into the malware itself."[10] In that world, significant destruction can be achieved with minimal human intervention. The logical recourse here is to boost investment in defensive AI capabilities as well, with the stakes in the spy-versus-spy game potentially becoming even higher as a result.

Supply chains are affected as well, as software in various imported devices may also be compromised. As a consequence, the U.S. government has taken some recent actions to protect the security and integrity of the bulk power system, including a 2020 Executive Order prohibiting acquisition and installation of any bulk power equipment that may have been supplied by an adversary of the United States.[11]

A further challenge with protecting the power grid in the coming years is that—with the adoption of solar panels, on-site batteries, electric car chargers, smart water heaters and air conditioners, and a host of other devices—the entity we need to protect is morphing into something new. In some ways, it's more vulnerable, while in other ways, it is becoming increasingly resilient. In the old world—only about a decade ago—protecting the power grid from cyberattack really meant protecting the isolated centralized SCADA systems for assailants. There was a physical fortress to protect, and the notion of "defense-in-depth" had real meaning. The term is frequently used to describe a layered approach to IT cybersecurity, and the concept was similar to that of the concentric protection strategies for a medieval castle. Attackers must first advance past a rain of arrows from archers on the parapets, then cross the moat and/or the heavily guarded drawbridge. Once inside, they still have to make their way past internal guard towers and other defenses to reach the coveted prize, which in yesterday's world would have been the SCADA system that was isolated from the outside environment.

However, in recent years, IT and ICS have converged to a large degree, even if they are separated by firewalls. Recently evolved "cyber-physical systems" now integrate physical grid assets with algorithms to create vastly improved situational awareness and greater grid efficiencies. However, these connected systems have also become vulnerable. A 2018 U.S. Department of Defense white paper on protecting cyber-physical systems in the energy and military sectors noted that a "single 'silver bullet'" will not stop all adversaries, and outlined the essential elements of an evolving defense-in-depth strategy.[12]

To get a better sense of the issue, I contacted Ray Rothrock, former CEO of RedSeal, in a July teleconference. RedSeal offers cybersecurity services to many industries, including utilities, and helps them inventory what they actually have on their system, where the vulnerabilities lie, and how to develop protection and resilience strategies. Rothrock summed up the main cyberchallenge: "The threat environment is the fact that people are automating everything. As people bring the digital network to touch the SCADA network, that's where the problems originate. The SCADA guys are pretty good. They have hardware and segmentation built in, and the equipment guys are pretty good. It's when you plug that TCP/IP cable in and interject the web server to extract digital data to run up to control systems, that's when you get into trouble." In other words, when you start connecting internet-networked devices with SCADA systems, you create potential vulnerabilities.

Rothrock further observed, "I don't worry about somebody opening or closing the wrong valve. I worry about somebody telling a legitimate operator that things are broken. It's the deception I worry most about. They (operators) take the job seriously and are well trained, and if they get signals they've got a problem they'll do something, using the logic against itself." He went on to say, "Probabilistic risk analysis is very effective at assessing equipment failures, but nobody has figured out how to quantify the human factor in complex systems. . . . This is a pervasive problem as digitization occurs in complex systems; the human factor just gets louder and louder for owners, operators, and consumers."

PROTECTING THE FORTRESS
WHEN THE FORTRESS IS EVERYWHERE

Clearly, a comprehensive stance to the security of critical bulk power assets—such as control rooms, central generating stations, and substations—should be employed, combining both appropriate systems and practices that focus well beyond mere protection of the perimeter. However, this approach still assumes that there *is* a relatively well-defined perimeter. Increasingly, that is no longer the case, as the grid has already rapidly evolved into something new and different.

Instead of the relatively limited number of devices and attack surfaces that have traditionally characterized SCADA systems, utilities now play host to rapidly growing populations of connected assets that today already number in the millions. Each of these "smart" and connected devices represents both a potential attack surface that could be targeted by hackers as a pathway into the utility, and also a potential weapon that can be used against the grid if its behavior is harnessed with intent to do harm.[13]

Many of these customer-sited, behind-the-meter solutions are specifically designed to interact either with the distribution grid or with the bulk power grid (even selling services into wholesale markets), or both. Some assets are simply designed to curtail consumption—such as thermostats programmed to use less energy by raising temperature set points during specific hours. Other devices, such as rooftop solar panels, inject surplus power into the grid. And an increasing number of devices—especially batteries to smart water heaters, and very soon, bidirectional vehicle-to-grid EV chargers—either intentionally absorb from or inject energy into the grid, creating increasingly large two-way power flows on a grid that was not designed for that behavior.

A GROWING POPULATION OF INTERCONNECTED DEVICES CREATES NEW VULNERABILITIES

Today, tens of thousands of households on Oahu have rooftop solar arrays, with the majority exporting electricity to the grid during midday sunshine, and others absorbing surplus energy in on-site battery systems.[14] The local utility, Hawaiian Electric (HECO), recently contracted solar installer Sun-Run to aggregate 1,000 residential battery systems for the utility to use as an aggregated resource.[15] HECO has also contracted with Shifted Energy to aggregate as many as 2,400 grid-integrated water heaters to create a 2.5 MW virtual power plant for HECO to dispatch when needed.[16]

There are two potential problems inherent in this scenario. The first is that many of these connected and aggregated devices are designed *precisely* to have a significant impact on the grid. That, in fact, is how they create value. And many of them reside at the distribution level, so the local impacts of a cyberattack could be quite significant. Batteries are integrated into the grid because they are able to absorb and release sizable volumes of electricity at moment's notice, so these would make an attractive target. The average size of battery installations going into homes right now is around 10 kWh, which is roughly equivalent to one-third of a household's average daily consumption. Aggregate enough of those to do the wrong thing, and you have a problem.

High-speed EV chargers may pose perhaps an even greater potential threat, when the networks are eventually built out. The fastest chargers for passenger vehicles can now flow 350 kW at any given moment, the equivalent instantaneous electricity draw of a decent-sized supermarket. Meanwhile, soon-to-be-deployed EV truck chargers will easily exceed 1 MW and are likely to be clustered at centralized locations the same way we see diesel pumps at highway truck stops today.[17] There are not enough of either type of chargers in concentrated locations for this to be a problem yet, but it is an issue that has gained the attention of security researchers at our national laboratories.[18] The thinking is that if hackers were able to manipulate these devices, they could potentially cause serious damage to both the connected vehicles and elements of the power grid.

Many of the most critical devices in this new energy ecosystem, especially batteries and solar panels, are connected by "smart" inverters to the grid. These little devices convert the DC to AC and automatically manage voltage and power quality. When functioning properly, they actually help maintain overall grid stability.[19] However, if hackers were to gain control of any sizable population of these devices, they could cause real damage and might even create blackouts.[20]

Then there are appliances like water heaters, washing machines, and refrigerators that will also be linked to the grid and change their behavior based on market or grid conditions. They will be aggregated and dispatched accordingly—told to stop or start doing what they are doing with an external signal. In fact, in September 2020 the Federal Energy Regulatory Commission (FERC) issued a ruling specifically allowing distributed energy resources to sell services to the wholesale power markets.[21]

The second critical issue is that many of the hundreds of vendors and solutions providers out there in the market do not place security high on their list of priorities. Many of these technologies and business models are new and constantly evolving, and in many cases driven by very young companies that are intent on demonstrating the feasibility of their technologies and business models while struggling to make payroll. In fact, one SCADA professional commented to me that he has observed solar companies "specifically placing inverters and protective relays on publicly accessible internet connections with default passwords and no firewalling, because it enables remote troubleshooting." Security issues are generally far down on these companies' lists of priorities, and investments in that area represent a cost center that will not yield a return.

The potential magnitude of this type of vulnerability was revealed in 2015 when one of the leading inverter players highlighted its efficiency in a blog, crowing about its ability to perform a remote firmware upgrade of 800,000 devices: "So the other day someone in a backroom in Enphase HQ quietly pressed the enter button and changed the settings on 800,000 inverters across 51,000 homes. No truck rolls. No field calls. No dogs to navigate. No chatty retired engineers to talk to."[22] This is just the sort of blog one hopes that hackers sitting in some office overseas are not paying attention to (but of course, they are).

CYBERATTACKS ARE NOT THE ONLY THREATS

Industry experts tend to describe widespread outages as either gray sky or black sky events. Gray sky events are outages that take place on a regional level—hurricanes and widespread ice storms are good examples. Such occurrences can usually be dealt with relatively quickly, often because utilities from across North America will quickly mobilize and hasten to aid the stricken party under long-established and effective mutual assurance programs. In a 2013 blizzard that crushed New England, I saw utility crews with their bucket trucks from as far away as Alabama, South Carolina, and Texas. The Texas crew told me that was an easy one, as they actually had a functioning hotel to sleep in. During Superstorm Sandy, they said they slept in their truck for five nights.

Far more devastating are the black sky events that can take down the grid and render it inoperable for weeks or even months. The severity of such an event would be magnified because other critical infrastructure—such as water, communications, natural gas, and food delivery—all rely on electricity to function. So if the grid goes down, everything else collapses along with it. That level of disruption would bring with it a high likelihood of social instability and potentially high levels of mortality. An effective and well-planned cyber event could potentially create a black sky event. But so too could an electromagnetic event—either natural or man-made—or a coordinated physical attack on critical infrastructure.

Solar Weather: A Chance of Coronal Mass Ejections

Every once in a while, the sun burps. When the sun's magnetic fields become contorted, they eventually snap rapidly back into realignment. When that occurs, they release enormous amounts of energy into space, in the form of either a solar flare or a more serious (to us) coronal mass ejection.

According to NASA, solar flares travel at the speed of light and reach Earth in eight minutes, accompanied by high energy particles that arrive somewhat later.[23] A flare's energy may affect communications signals temporarily, but effects are generally minimal. By contrast, a coronal mass ejection (which sometimes accompanies the largest flares) involves a concentrated and directed cloud of magnetized particles—plasma—moving a million miles an hour and taking up to three days to reach us.[24] A coronal mass ejection can disrupt the Earth's magnetic fields (creating an aurora, among other things), and create electric currents that can overload power grids.

These events tend to concentrate and amplify along extended power lines, and the March 1989 solar storm that centered in Quebec provides an example of the potential impacts. That resulting geomagnetic disturbance took out the provincial power grid for almost nine hours.[25] However, the damage was not limited to Quebec. Over the following day, effects were noticed by many utilities to the south, and a later analysis by the North American Electric Reliability Corporation detailed 211 "Reported Events" recorded by utilities across the continent.[26] A subsequent paper on the event observed that much of the damage was not noticed for some time, since it was not always immediately evident. As an example, 11 nuclear plants reported large transformer failures within two years of the disturbance.[27]

The 1989 solar storm is not even a worst case. In August 1859, a very large solar storm (known as the Carrington Event) slammed into the Earth and took out telegraph lines around the world, with some catching fire. In Washington, DC, an operator suffered a shock when an arc of fire leaped from the equipment to his head.[28] The Carrington Event had an estimated disturbance index of −850 nanoteslas (nT), compared with the Quebec event that was measured at −589 nT.[29] It might be, though, that neither event represents the limit. NASA reported in July 2012 that one of its orbiting satellites—with equipment on board to measure the event—took a direct hit from a coronal mass ejection, with an estimated Dst index of −1,200 nT. That phenomenon ripped straight through the Earth's orbital path, but fortunately we were elsewhere on our trajectory through space.

One of the scientists who authored a 2013 study on the storm commented, "If it had hit, we would still be picking up the pieces. . . . If the eruption had occurred only one week earlier, Earth would have been in the line of fire."[30] The report also indicated that analysts believe that such an event would be capable of "disabling everything that plugs into a wall socket."

Electromagnetic Pulse Attacks

Electromagnetic Pulses (EMPs) are yet another existential threat. The potential impact of EMPs on the electric grid first became apparent in 1962, when the U.S. military detonated a 1.4 megaton nuclear bomb 240 miles above the earth, 900 miles away from Hawaii.[31] The EMP that resulted blew out streetlights, and disrupted radio and telephone communications. It became clear that an EMP could destroy the utility computer and SCADA systems that oversee the electric grid.

While the U.S. military made efforts to harden its infrastructure in response, he power grid has remained largely unprotected.[32] In an attempt to remedy this deficiency, Congress commissioned a report in 2004, followed by the more detailed 2008 "Report of the Commission to Assess the Threat to the United States from Electromagnetic Pulse (EMP) Attack." The latter report commented that, "A single EMP attack may seriously degrade or shut down a large part of the electric power grid in the geographic area of EMP exposure effective instantaneously." It further elaborated, "There is also the possibility of functional collapse of grids beyond the exposed one, as electrical effects propagate from one region to another. . . . Should significant parts of the electric power infrastructure be lost for any substantial period of time, the Commission believes that the consequences are likely to be catastrophic, and many people may ultimately die for lack of the basic elements necessary to sustain life in dense urban and suburban communities."[33]

A 2019 document on the topic from the North America Electric Reliability Corporation (NERC) reaffirmed the notion that an EMP could broadly affect the U.S. power grid.[34] To provide a sense of the potential impact, an EMP explosion 30 miles overhead could impact a radius of nearly 500 miles. A 300-mile-high detonation (from a missile or satellite) would create a 1,500-mile radius, which would affect most of North America. The potential enemies with missiles capable of causing that level of destruction include China, Russia, Iran, and North Korea, which has indicated that the ability to launch a "super powerful EMP attack" is a strategic military objective.[35]

FERC and the Department of Homeland Security (DHS) offer limited guidance as to how to respond to the threat, but in general, preparations have been minimal and the overall policy response tepid. Today there is neither a clear game plan for addressing this situation, nor clarity concerning who would pay for the necessary investments.

In March 2019, the Trump administration issued an "Executive Order on Coordinating National Resilience to Electromagnetic Pulses."[36] It directed affected government agencies to address the issue of both human-induced and naturally occurring EMPs. DHS is to "use the results of risk assessments to better understand and enhance resilience to the effects of EMPs across all critical infrastructure sectors, including coordinating the identification of national critical functions and the prioritization of associated critical infrastructure at greatest risk to the effects of EMPs." It also called upon the Secretary of Energy "to conduct early-stage R&D, develop pilot programs, and partner with other agencies and the private sector, as appropriate, to characterize sources of EMPs and their couplings to the electric power grid and its subcomponents, understand associated potential failure modes for the energy sector, and coordinate preparedness and mitigation measures with energy sector partners."

Today, this country is woefully unprepared to address the potential of an EMP event, either man-made or naturally occurring. Some have argued that an EMP attack from an adversary is unlikely, since the United States could respond to any aggressor with significant military force. However, that doesn't address solar flares and coronal mass ejections. Threat of military retaliation would be far less persuasive when directed at the sun.

Physical Threats

In the middle of the night on April 16, 2013, just outside of San Jose, California, a small number of individuals quietly approached Pacific Gas & Electric's Metcalf substation. They cut the fiber-optic communications line that allowed the facility to communicate with the broader network. In the next 20 minutes or so, the unknown assailants fired over 100 rounds of ammunition into 17 transformers, causing the cooling oil to leak and taking the transformers and substation out of service.[37] Fortunately, the utility avoided outages by rerouting power through other parts of its transmission infrastructure.[38]

This event unnerved observers, who initially thought it might be a dress rehearsal for some larger event. A coordinated assault on a number of the country's most critical high-voltage transformers could be much more serious, since as much as 60%–70% of the nation's electricity moves over an estimated 3% of the U.S. transformers.[39] They are often custom-built overseas, taking between 5 and 16 months to manufacture, and many are not interchangeable.[40] Transformers are large and extremely heavy, with some weighing in at over 400 tons. Thus, if an enemy were to destroy a critical number of them, serious economic damage and potential long-term and widespread power outages could result.

A number of programs have been put in place to meet this challenge, with some stockpiling of transformers taking place in recent years, and plans among utilities to mutually support one another in the event of transformer losses for whatever reason. The shortcoming here is that mutual assurance programs can become stressed or fail when an event is widespread, and numerous utilities are clamoring for the same limited resources available in the stockpile.

THE VALUE OF PLANNING AND THE NEED FOR COORDINATION AND INVESTMENT

Our ongoing experience with COVID-19 has been devastating. And yet as bad as it has been, it could have been far worse in terms of its mortality, disruption to civil society, and even its effects on the power grid. Nonetheless, it has created stresses in the utility industry, including the need for control room area personnel in some regions to go into full lockdown mode, isolated from their families for months on end in order to assure continuous operation of the grid. At the same time, the response of the New York ISO grid operator (NYISO), in the state most immediately affected by the pandemic in the first days of the unfolding disaster, demonstrated the value of planning and of developing useful playbooks. When COVID-19 came, the NYISO operator followed its previously scripted plan, with some alterations.[41] Other grid operators across the country took similar precautions.

This experience serves as a reminder as to why it is necessary to create the plans and make the critical investments well ahead of any potentiality. The challenge of safely managing a control room in a pandemic is not insignificant, but it pales in comparison to the planning and response needed for the other, region-wide, threats. Many of these will require plans and related investments that are far more complex and significant.

To date, although we have been aware of these threats for decades, we as a society have taken little meaningful action—and made only relatively limited investments—to address them.[42] Our current approach reflects the unfortunate weakness of our society, which is (both understandably and regrettably) less focused on the potential magnitude of the damage and too focused on the potential economic costs and who will bear them. This is a classic market failure, where the potential external costs have not been baked into the economic equation and nobody is paying to avoid the potential negative outcomes. It's also a case for which the federal government must step in, change the current approach, and ensure that the requisite investments are made, using U.S. taxpayer dollars if necessary.

For each of the threats described above, there are specific actions we can take. They include a larger number of drills and preparatory exercises, development of playbooks for numerous contingencies, and more formalized sharing of information between various parties.[43] A December 2018 report from the President's Natural Infrastructure Advisory Council (NIASC), "Surviving a Catastrophic Power Outage," summed up our woefully inadequate national preparedness in just a few words: "After interviews with dozens of senior leaders and experts and an extensive review of studies and statutes, we found that existing national plans, response resources, and coordination strategies would be outmatched by a catastrophic power outage."[44]

Whether we like it or not, the next battlefield may indeed take place where we live. COVID-19 provides ample warning of what happens when an unwelcome Black Swan arrives on our shores. We cannot say we have not been warned.

15

NAVIGATING
TOWARD THE FUTURE

In the past few years, an increasing number of cities, eleven states, and yes, even six utilities as of mid-2020, have publicly committed to 100% clean energy (with the targeted date generally between 2040 and 2050). One in three Americans now live in a city or state committed to 100% clean electricity.[1] Many more will follow as the climate threat becomes even clearer, but that transition will not always be easy, and the roadmap will not always be clear. It will involve cultural shifts, a workforce with new and different skills, and courageous leadership willing to make difficult trade-offs and embrace uncomfortable change.

Mary Powell, Green Mountain Power's former CEO, is known as a leader who did just that. Among other accomplishments, she turned around the struggling utility,[2] which eventually led to Green Mountain becoming the first utility with B Corporation certification, specifically committed to meeting certain social, environmental, transparency, and accountability standards.[3] Powell firmly believes the way to accelerate that desired transition is to change the underlying culture. She emphasized to me in a recent conversation that to successfully address these looming challenges, it is critical to bring a sense of hope, possibility, and even abundance to the discussion. If one thinks about the issue that way, and embraces a sense of possibility, she commented, "You are way more productive, and you can get so much more accomplished. A way to look at the glass as half full—when the planet is on fire—is that it has presented an abundance of opportunities to be part of building the solutions."

In looking forward, then, and trying to project into the future based on what we know today, how might we get to the clean grid of tomorrow? What trade-offs and decisions will we have to make? And how might we even more proactively employ clean electricity to decarbonize the broader global energy economy?

GETTING TO 100% CLEAN POWER

How do we get to that goal? Like the old saying about how to go about eating an elephant—you do it in pieces—it probably helps to conceptualize this by breaking it into phases.

To consider the process of transitioning away from a "traditional" grid characterized by fossil-fuel resources such as coal- and gas-fired plants, let's adopt the software language many of us are familiar with—the concept of a version 1.0, 2.0, 3.0, and 4.0. Grid 1.0 would be the traditional power grid prior to the introduction of renewables, depending largely on fossil resources, with some existing hydropower also in the mix. One can quibble about the percentages, but Grid 2.0 would involve the integration of renewables to get to a roughly 30% decarbonized grid. Grid 3.0 would include somewhere on the order of 60 to 80% carbon-free resources, and Grid 4.0 would be entirely carbon-free.

Some 3.0 power grids already exist, such as those of Iceland and Québec, but they were blessed with abundant low-cost hydroelectric resources to start with.[4] They never had to evolve through a deliberate transition, and they don't offer much in the way of lessons for us. Therefore, let's leave them aside and focus on the majority of our carbon-dependent grids.

Grid 2.0: Existing Grid Resources Accommodate a Growing Portfolio of Renewables

Not so long ago, planners and skeptics believed it would be difficult to build enough solar and wind resources—and even more challenging to integrate their intermittent output—in order to arrive at a 30% decarbonized grid. In fact, many planners thought a couple of decades ago that it would be difficult to integrate just 5% wind and solar resources into the mix without incurring increased costs.[5] The intermittent nature of the wind and solar resources was believed to be too difficult to successfully integrate.

Integrating large quantities of these renewables meant that other resources had to change their operating behaviors since wind and solar power plants could not be "dispatched" by the grid operator to match constantly varying demand (known as following load). Power plants generally have a reliable fuel supply—gas in a pipeline or coal in a pile—and they are generally pretty straightforward to dispatch. By contrast, renewables literally show up depending on when the wind is blowing or when the sun is shining.

Despite that challenge, a number of grids have demonstrated that large quantities of wind and solar energy can be accommodated. In part, this was accomplished by reconsidering how existing "dispatchable" assets can behave and forcing generating plants—especially the gas-fired fleet—to become far more flexible, ramping up and down (cycling) to accommodate the flux in renewable resources. It also involved better forecasting tools, allowing operators to better predict how much energy was likely to be produced in the near future, so they could more efficiently operate the other resources in the system. A broader transmission network is also helpful, since it allows the movement of surplus renewables to neighboring grids.

Grid 2.0 is characterized by two key concepts: 1) the grid accommodates all of the renewable resources possible, with a goal of not wasting (curtailing) solar and wind unless other fossil plants cannot be backed down any further, and additional renewable output simply cannot be accommodated; and 2) existing generation equipment behaves more flexibly than in the past (even though it can lead to premature aging of the equipment). There is also some existing demand response, where customers cut consumption in response to a grid signal. In a Grid 2.0 model, utility-scale energy storage is not yet a part of the mix. Furthermore, the customer side of the equation or "grid edge" is relatively underdeveloped. There may be some rooftop or customer-sited solar and a few batteries, but these are not yet aggregated and coordinated to become a directed grid resource.

COVID-19 gave us a glimpse into the future and helped show us that this level of renewables integration could actually be achieved without much difficulty. As a result of the economic lockdown, electricity demand curves across the country fell, and supply declined along with it. The interesting part of that dynamic was as the fossil-fired resources were cut back, the absolute level of renewable generation stayed the same, so that its overall contribution as a percentage of the total increased. In some areas, the renewables mix jumped to as much as 40% for a short period of time during the pandemic, and for the most part, those resources were integrated without difficulty.[6]

Grid 3.0: Storage Is Added to the Growing Mix of Renewable Assets

In a Grid 3.0 model, which is more difficult to achieve, a number of developments occur, and the planning and integration effort must be much more intentional. The most critical aspect of Grid 3.0 is that *the job of the rest of the assets on the grid is to essentially wrap around and accommodate as many of these clean and increasingly low-cost electrons as possible.* Energy storage and a growing population of responsive customer-based assets are brought into the game. A large infusion of and some also become necessary to manage a vast network of connected, constantly flexing assets. To create a grid characterized by 60%–80% carbon-free assets, the following strategies are likely to be employed:

1. Existing nuclear plants will be maintained and operated for as long as possible, so long as they remain cost-effective to operate.[7]

2. A much higher level of low-cost wind and solar resources will be built and connected to the grid as renewable resource costs continue to fall inexorably with scale and improved technologies (offsetting the loss of any subsidies that exist today). Solar generation assets, in particular, will be purposely overbuilt. This will allow them to operate with more flexibility and ramp up and down their production as needed by grid operators. Solar panels will be so cheap, it won't matter if some of the potential solar output is curtailed—the added flexibility that solar brings to the grid will more than make up for it.[8]

3. Energy storage (largely in the form of batteries) will increasingly support both wind and solar resources as a default option, allowing owners and operators to shift resource output, generally by four hours. This will vastly increase the amount of renewable energy that can be integrated.

4. The population of distributed energy resources (DERs) on the customer side of the meter—so-called grid edge devices such as batteries, controlled thermostats, EV chargers, and water heaters that can be proactively and remotely manipulated—will grow rapidly (the recent FERC Order 2222 specifically promotes this population of devices and calls for their interaction with power markets).[9]

5. An increasing number of EVs will be capable of both charging at opportune times and discharging into the grid when needed.

This level of increased flexibility across the entire landscape of generating and storage assets, and customer devices, will further allow for more increasingly low-cost renewables to become integrated into the grid.

Grid 3.0 is feasible today in many parts of the country with existing technologies, though various elements of this scenario will require some level of subsidy if we are to accelerate this process. The Grid 3.0 scenario will initially be developed in areas where the costs of electricity are the highest—one reason why Hawaii is so far ahead of many other states in renewables deployment and integration. It's far more cost-effective to put solar on your roof when the cost of the electricity you avoid buying is over 30 cents per kWh, compared with 10 or 20 cents in other parts of the country.

The Kaua'i Island Utility Cooperative (KIUC) points the way to a 3.0 model, and is explicitly committed to having solar, biomass, and hydropower produce at least 70% of the island's electricity by 2030. In 2019, it generated 56% of its electricity from carbon-free resources. In 2020, on any given day, all the electrons generated and sold for at least five hours per day, were entirely carbon-free.[10]

KIUC made that transition from 8% to 56% renewables in only 10 years, by installing 64 MW of utility-scale solar projects combined with 32.2 MW of rooftop solar on people's homes. And then it added a liberal dose of energy storage. In its first 13 MW solar project, installed in 2017, Kauai added a 13 MW battery capable of delivering four hours of storage, which helped meet evening power consumption when the sun set and demand was high. Two years later, another 20 MW of solar went in, coupled with a battery of the same size, which together provide the ability to shift 100 MWh to address the evening peak demand.[11]

KIUC depends heavily on solar energy, but different utilities and grid operators will approach this challenge differently, in large part based on the regional resources available to them. The Pacific Northwest, for example, will need fewer batteries as it can take advantage of its large hydro resources. The American Southwest will rely largely on solar energy, with some nighttime wind to complement the sun's energy and enough storage to buttress the entire mix and make it somewhat dispatchable. The U.S. Midwest will rely more heavily on wind energy, and perhaps transmission to even out the associated intermittency. Meanwhile, much of the northeastern United States will likely lean significantly on offshore wind as well as a mixed bag of other resources, and the southeastern states will be heavily solar-dependent. In one way, shape, or form, though, each region should reasonably be able to achieve that 60%–80% carbon-free threshold with existing technologies.

The Reality of Making That Transition

Conceptually, this looks easy. In practice, it is anything but. There's a fine balancing act for utilities facing the pressures to reduce their emissions while balancing costs and responsibilities to customers. To get a better sense of the specific challenges of walking that tightrope, I spoke with CPS Energy's president and CEO, Paula Gold-Williams, on a teleconference call one summer afternoon in 2020. CPS Energy—owned by the City of San Antonio—is the largest publicly owned electric and natural gas utility in the country and faces a city mandate of being 100% carbon neutral by 2050. Gold-Williams is a highly regarded thought leader in the industry because of her ability to effectively conceptualize and communicate those challenges and difficult trade-offs and motivate her team.

She observed that the transition "is not linear," because both customer needs and technologies continue to evolve. San Antonio, Gold-Williams commented, "is one of the poorest large cities in the country," so CPS Energy must be very thoughtful about the costs it incurs, which are eventually borne by the city's residents.

She reminded me, "We are a utility company and we are not legislators. So, we don't lead with policy. Now our city has set goals—a Climate Action and Adaptation Plan—but created these goals that say we want to get to net-zero carbon. . . . We want to inform that conversation. We want to talk about the technological changes and what we need to do. We are at the table informing and talking about the pros and cons of technology."

Faced with that constantly shifting landscape of changing technologies, CPS Energy created an approach it labels the "Flexible Path^{SM}," which calls for the utility to meet its goals with a high degree of adaptability and optionality. As part of its journey, the utility closed two coal-fired plants in 2018 and increased its renewables supply from 13% to 22% in the 2010–2018 period, adding substantial amounts of wind and solar resources.[12]

In our conversation, Gold-Williams focused on the issue of carbon reduction versus the economic trade-offs, noting that it's not like one can just snap a finger and effect the transition. For example, she said, "We still have some younger coal units left, and it's very challenging to simply junk those and transition to something new"—especially when the utility has bonds to pay. The challenge in the carbon conversation is not about the goal, she said, "but the velocity of change. Some people say you have to just jump off (fossil fuels), and then that will solve all your problems, but those are the people who don't have the obligation to supply the power. But if we do it in a way that's thoughtful over time, we will get to where we are supposed to be." Gold-Williams noted that CPS Energy already has 1,000 MW of wind and 600 MW of solar, and the utility has gained a good deal of experience in "managing operational challenges, smoothing out cost implications, and making plans for the future."

She emphasized that it's important to think about the demand side of the equation as well, since investments in energy efficiency technologies are often more cost-effective and can be targeted across many demographics in the city. She cited the Save for Tomorrow Energy Plan (STEP), an efficiency program that saved 800 MW of peak demand, which helped CPS avoid the need for investments in additional generating capacity some years back. "If we hadn't saved 800 MW, we would have built a large combined cycle plant which would probably be five years old at the most, and we would be locked into that technology."

Efficiency investments have bought CPS Energy valuable flexibility in a period of remarkably fast change, Gold-Williams said. "Anything you can do to buy time and knowledge is what you want to do. Things are happening in one year's time or two years' time. We have leveraged our efficiency and conservation program. It has allowed us the leverage to pump up our knowledge and technology views."

Gold-Williams also discussed the challenge of changing the behavior of the existing assets to accommodate more renewables, how to blend those assets into the mix, and how to navigate the path to a carbon-free grid. "It's easier said than done depending on how old your plants are. That requires a lot of thought on the transition and how you will evolve. You put yourself on a path that says 'I think I can bite off a 20% transition in the next four years, and then another 20%.' So, your general plan evolves from moving into a 50-year asset, to a focus on shorter intervals and getting investment levels lower. And the life of the new asset is shorter, because you expect the technology to keep going. If you apply Moore's Law[13] and accept that it will come into the utility industry, whether in energy production systems, or even in the back office, you're always searching for a better deal."

The next step for CPS Energy is its forthcoming global Request for Information for their *Flex*POWER Bundle[SM] initiative that aims for up to 900 MW of solar, 50 MW of storage, and up to 500 MW of some firming capacity that would involve new technology solutions.[14] "I am so excited about the opportunity for people to tell us things we need to know. The economics have improved, and we want to see that," said Gold-Williams.

At the end of the day, she continued, "You have to think of everything in terms of multiple choices and pick the best one. Will you always make the best pick? No—but you've got to make the best choice and try not to cause your customer economic shock. And you have to think about stranded costs of any technology you are retiring because your customer is still paying for that even as you are moving onto something else."

Most U.S. grids are in the transition to a 2.0 mode today, with some version of 3.0 on their planning horizons. Most also face an increasing volume of renewables pouring into the system, a conventional generation fleet that is being challenged to respond much more flexibly, and few, if any, new storage resources. A number of states—such as California, Hawaii, Massachusetts, New York, Texas, and several in the windy Midwest—will soon close in on that 30% level within the next few years, and they won't stop there. As Gold-Williams made it clear, the further one gets down the road, the value of flexibility becomes critically important, as even more trade-offs must be made.

Grid 4.0: The Last Mile Becomes Even More Difficult

From there, the challenge of moving to the 100% carbon-free grid becomes even more daunting. Lengthy transmission lines can help a great deal in zipping streams of renewable electrons across the country from where they are generated to markets hungry for clean power.

Other smart DER on the customer side of the meter will pitch in as well, with everything becoming smarter, connected, and more responsive to grid conditions and prices. For example, a well-orchestrated fleet of connected electric vehicles charging and discharging at opportune times could add a great deal of value to the grid. So, too, could millions of smart, efficient water heaters and air conditioners. The size of this potential DER reservoir may be enormous, with estimates ranging from 65,000 MW to 380,000 MW by 2025[15] (to put those numbers in perspective, the country currently has approximately 1,100,546 MW of generating capacity.)[16]

Nonetheless, even with the potential emergence of such flexible resources, a 100% carbon-free goal is a daunting one, and it's not easy to see from today's vantage point exactly how we will get there. But before the notion of a carbon-free grid is discarded as an impractical future impossibility, it's worth bearing in mind just how quickly the pace of change has occurred to date: the wind and solar industries have only existed for a decade at any meaningful scale, and batteries and other new energy storage technologies are just getting started. Such technologies can enter the landscape with astonishing speed, if the economic value is there. And some of those may be just around the corner.

FUTURE RESOURCES
THAT MAY HELP AID THE TRANSITION

Fusion

Although it has continued to remain elusive for decades, fusion is one such possibility. For many years, the standing joke has been that fusion is the energy source of the future and always will be. Many of the fusion-related projects have been extremely large government-sponsored initiatives, but that scale may be the element that has inhibited the ability to innovate and make the necessary changes within timeframes that can keep up with the competing technologies out there.

One company, Lockheed Martin, has been quietly moving forward on a smaller-scale approach to nuclear fusion that it believes may hold promise. In 2014, Lockheed announced that its Skunk Works team was working on a compact fusion reactor that could be "developed and deployed in as little as ten years." The key critical concept was to work on smaller machines so that it could iterate—design, build, test—much more quickly, thereby accelerating the process of learning. The company announced it expected to have a prototype within five years.[17]

The company has largely gone silent in the intervening years, but apparently developed four different reactor test designs in the interim, and is now at work on a large and more powerful fifth iteration. The critical question with fusion is how to handle the heat—temperatures reaching hundreds of millions of degrees—and pressure created in the process of fusion, the same process that occurs in stars.[18] If Lockheed can address the challenge and create a cost-effective small fusion device, it could indeed change everything. But there's a long way to go, and aside from the technical issues, there is always the overriding question of costs.

Meanwhile, scientists at the Massachusetts Institute of Technology's Plasma Science and Fusion Center tout their work to start building a new fusion device in 2021 that will create a self-sustaining fusion reaction in a controlled manner. This effort is seen as the precursor to a future working prototype of a fusion-driven power plant.[19]

Large Centralized Nuclear Plants

One technology going nowhere fast is large-scale nuclear power generation. The last three attempts here in the United States have all culminated in financial disasters. The costs of the proposed Levy plant in Florida soared from an original price tag of either $3.5 billion or $6 billion to $22.5 billion[20] before it was canceled. Ratepayers paid over $1 billion in planning costs, but at least there was no construction work undertaken that would have greatly magnified the bill.[21]

Ratepayers in South Carolina suffered a far worse fate, with the ill-fated V. C. Summer plant incurring $9 billion in costs before it was terminated, leaving perhaps the world's most expensive hole in the ground.[22] Ratepayers have already paid $2 billion for the project and will get nothing in return.[23]

That leaves only one utility-scale project left under construction in this country, Georgia's (now) $27 billion Vogtle plant. This plant—originally budgeted at $14.3 billion—has suffered similar overruns and extended timelines, and led to the bankruptcy of Westinghouse Electric, but state regulators continued to support its construction.[24] In 2020, independent monitors released a report suggesting an additional $1 billion in cost overruns and further delays were likely, and commented on an "astounding 80%" failure rate related to new components installed at the power plant.[25]

Such experiences related to large-scale nuclear construction are by no means limited to the United States. Nuclear facilities in France, Finland, and the United Kingdom have been plagued by similar delays and cost overruns. A critical shortcoming is that building a large nuclear reactor is like building a very complex stick-built home. Each aspect of the process is both bespoke and linear, so one unanticipated delay holds up all downstream scheduled activities; all the while the clock is ticking, and costs continue to mount. Based on this mounting body of evidence, it is safe to say that we will probably never see another large-scale nuclear reactor built in this country.

A More Modular Approach Might Work

That does not mean, however, nuclear energy is necessarily destined to fade away, as another more flexible and cost-effective approach may be in the offing. Small nuclear modular reactors (SMRs) may be the solution.

The concept here is simple: The SMR approach facilitates construction of various parts on an assembly line. These elements can be transported in pieces by truck or rail and assembled on-site. It's the modular home construction model on steroids, and offers significant potential because this cookie-cutter approach allows for rigorous quality control within the factory, relatively easy transportation of the various elements, and a standardized on-site installation. The modular aspect offers more flexibility for investors and power companies looking to meet demand. They also don't require a large physical footprint.

Among the companies competing in this space, NuScale Power is perhaps the farthest along. It has a 60 MW light water reactor with a safe design that ensures that it cannot melt down. It can also be daisy-chained together with as many as 12 units comprising a single power plant.[26] NuScale is currently working on a 720 MW plant for the Utah Associated Municipal Power Systems, with commissioning of the first module expected in 2029.[27]

Here's the big potential problem, though: NuScale puts the targeted levelized cost of producing the energy (meaning all-in costs, including capital, interest, operations and maintenance, and decommissioning) at $65 per MWh.[28] By contrast, NextEra Energy—the largest developer of renewables in the world—expects the unsubsidized levelized cost of wind combined with four hours of battery storage to be between $20 and $30 by 2024, while solar plus storage comes in at between $30 and $40.[29] NuScale's $65/ MWh goal is also roughly twice the average wholesale price of electricity in U.S. competitive markets in recent years.

That's a potentially serious problem for NuScale and its nuclear competitors. SMRs do have the advantage of being able to produce energy around the clock, while renewables, even with four hours of storage, are more limited. However, solar panels, wind turbines, and batteries all continue to see both improvements in technology and consistently declining costs associated with production volumes. NuScale and others in the SMR space may simply find themselves in a global race against other technologies enjoying hundreds of billions of dollars of investment and R&D capital, a race they may have difficulty winning unless they can scale production, cut costs, and deliver energy at a far lower price tag.[30]

Hydrogen

Then there's hydrogen. The little H_2 molecule could potentially be an industry game-changer, for a couple of reasons. First, it can be produced with renewable energy (green hydrogen), creating a carbon-free fuel. Second, when you burn or use hydrogen in a fuel cell, the by-product is water.[31] Hydrogen is therefore both a way of storing renewable energy and using it as fuel for multiple applications.

As storage, hydrogen represents that potential holy grail of a way to capture and store electricity in bulk for long-term use during periods when the wind doesn't blow or the sun doesn't shine for extended periods.

As raw fuel for combustion, it may be the way to feed our power plants in the future so they are entirely carbon-free and can be dispatched to meet demand whenever needed. In fact, the first green hydrogen-sourced power plant is already being planned by the Los Angeles Department of Water and Power. The 840 MW facility is scheduled to come online by 2025 and will start burning a mixture of 30% hydrogen and 70% natural gas, eventually switching to 100% hydrogen.[32]

Fuel cells can also be used to convert hydrogen back to electricity, an application that may be particularly valuable in urban areas where conventional power plants and their associated emissions are unwanted and difficult to site. On-site fuel cells could solve that problem. In the transportation sector, hydrogen fuel cells may also play a role.

Hydrogen can also be used as a fuel in heavy industrial applications—such as the manufacturing of steel—in place of coal, coke, or natural gas. In short, the production of green hydrogen from renewable energy resources may allow society to use clean electricity to address climate change well beyond the electricity sector. Hydrogen's broad applicability across multiple sectors could help accelerate investment in the hydrogen space and grow the economies of scale. However, before we will see widespread application of hydrogen throughout the economy, a few fundamental issues remain to be solved.

The most critical issue is cost: of the electricity as an input, of the basic equipment necessary to create the gas, and of the infrastructure costs to move and store the gas. The production of green hydrogen involves the process of electrolysis—the separation of water into hydrogen and oxygen by means of applying an electric current to water in an electrolyzer. For green hydrogen to be able to compete with "blue hydrogen" (derived from fossil fuels through a steam reformation process with carbon capture and storage technology on the back end), both the cost of the renewable electricity and the cost of the electrolyzers must fall considerably. The good news is these cost declines are already occurring, but there is some way to go before green hydrogen will be cost-competitive.[33]

The costs of wind and solar projects are falling rapidly enough so in some cases that developers are now planning to overbuild projects and curtail energy at times when the grid cannot absorb it. That curtailed and otherwise useless energy may represent an opportunity for renewables as a cheap feedstock. However, if renewables are to do the job of feeding a hydrogen economy, they will have to be built out at a nearly unfathomable scale.

We are also beginning to see some announcements in very large renewables-to-hydrogen projects. Developers—largely using European offshore wind—are planning massive projects entirely devoted to the production of hydrogen. For example, in February 2020 Royal Dutch Shell and Dutch gas company GasUnie announced plans for an offshore wind-to-hydrogen project starting in 2027 that could grow to 10,000 MW by 2040.[34] They were joined by Norwegian and German energy giants Equinor and RWE in December.[35] In May, Danish offshore wind leader Ørsted also announced its plans to work with a consortium of companies on a 5 GW wind-to-hydrogen project.[36]

Hydrogen from renewables is not limited to Europe. Australia and the Middle East are also considering large projects.[37] In the United States, renewables developer NextEra announced in 2020 a plan to invest in its first green hydrogen plant, using electricity from solar energy that would otherwise have been wasted.[38] Over the course of the coming decade, we will have a much clearer picture of the viability of hydrogen as an energy carrier and its potential role in addressing the carbon challenge. For now, hopes are high, and tens of billions of dollars are being dedicated to addressing the opportunity.

2050 MAY FEEL LIKE A LONG TIME AWAY, BUT THE URGENCY IS NOW

All of these technologies are in play, and more will likely come to the fore as humanity continues along a journey we have never attempted before: a global transition to a clean energy economy. While different countries face the same problem with varying levels of urgency, the need for successful outcomes is clear. Only a decade ago, it seemed this transition would not be feasible without a great deal of economic sacrifice. To many, the potential to largely decarbonize our global economy seemed unachievable.

In just 10 years, though, we have developed a continuously expanding and cost-effective toolkit that gives us hope that we have a reasonable chance of achieving that goal. With more powerful computers, available and willing capital, and clean energy technologies, for the first time we see the clear possibility of creating a global energy economy that is better than the one we hope to leave behind: cleaner, more resilient, and more cost-effective. We don't yet have all the tools we need, but perhaps that doesn't matter. Those will come with time, and perhaps sooner than we think.

Nonetheless, even the best tools on the planet are only useful if we possess the creativity, commitment, and courage to move forward on a journey we have no choice but to take. The real question is not when we achieve the carbon-free economy, but how long it will take us to get there, and the degree of suffering we can avoid if we accelerate that process.

We must continue to develop new business models, enhance the regulatory and political discussions, and effectively apply new technologies as soon as we are able to do so. We must embrace a sense of what is achievable and—as Mary Powell reminds us—do so with a belief in the possibility of the abundance that we are capable of creating, even if our planet is on fire.

As CPS Energy's Paula Gold-Williams put it, "Now things are changing every couple of years, or within a year. You can't feel it because you're in it, like you can't feel the earth spinning because you're on it. . . . There's still a good portion of people that say 'but do we really have to change?' I spend time telling them everything's going to change."

Indeed, everything will change. It will likely evolve much sooner than we think, and in ways we cannot foresee. Welcome to a brave new world.

NOTES

CHAPTER 1

1. Michael Holtzman, "Shuttered Brayton Point Assessed Value Drops More Than 75 percent," *Herald News* (Fall River, MA), November 21, 2018, https://www.heraldnews.com/news/20181121/shuttered-brayton-point-assessed-value-drops-more-than-75-percent.

2. "Brayton Point Power Station in Somerset, Largest Coal-Fired Plant in New England, to Close by 2017," Masslive.com, October 8, 2013, https://www.masslive.com/news/2013/10/brayton_point_power_station_in.html.

3. "Brayton Point Power Station," Sourcewatch.org, last edited January 15, 2020, https://www.sourcewatch.org/index.php/Brayton_Point_power_station.

4. Before 2012, the plant's cooling system also circulated approximately 1 billion gallons of water daily, releasing superheated water into the bay and imperiling hundreds of billions of fish larvae over decades, with resulting declines in various fish species. The cooling towers and a closed-loop cooling system were built in 2012 to solve that problem. Unfortunately, those investments coincided with the advent of new gas fracking technology and resulting in gas-fired power plants that could deliver energy to market far more cheaply.

5. Holtzman, "Shuttered Brayton Point Assessed Value Drops More Than 75 Percent."

6. In June 2020, for example, two utilities (Tucson Electric Power and Colorado Springs Utilities) each indicated that they were going to retire coal plants and construct new renewable resources without adding any gas plants to ease the transition. See Dennis Wamsted, "IEEFA U.S.: Utilities Are Now Skipping the Gas 'Bridge' in Transition from Coal to Renewables," Institute for Energy Economics and Financial Analysis, July 1, 2020, https://ieefa.org/ieefa-u-s-utilities-are-now-skipping-the-gas-bridge-in-transition-from-coal-to-renewables/.

7. Eanna Kelly, "COVID-19 Has Shown the Huge Cost of Slow Political Action, Warn Climate Scientists," Science Business, July 2, 2020, https://sci-

encebusiness.net/covid-19/news/covid-19-has-shown-huge-cost-slow-political-action-warn-climate-scientists.

8. "U.S. Coal Consumption in 2018 Expected to Be the Lowest in 39 Years," U.S. Energy Information Administration (hereafter abbreviated EIA), December 28, 2018, https://www.eia.gov/todayinenergy/detail.php?id=37817.

9. "NextEra Energy Inc. (NEE) Q1 2020 Earnings Call Transcript," Motley Fool, April 22, 2020, https://www.fool.com/earnings/call-transcripts/2020/04/22/nextera-energy-inc-nee-q1-2020-earnings-call-trans.aspx.

10. Zachary Shahan, "Tesla Model 3 = 7th Best Selling Car in USA," Clean Technica, January 19, 2020, https://cleantechnica.com/2020/01/19/tesla-model-3-7th-best-selling-car-in-usa/.

11. "June 2020 Tied as Earth's 3rd Hottest on Record," National Oceanic and Atmospheric Administration (hereafter abbreviated NOAA), July 13, 2020, https://www.noaa.gov/news/june-2020-tied-as-earth-s-3rd-hottest-on-record.

12. "Global Carbon Budget 2019: Emissions," Global Carbon Project, accessed November 10, 2020, https://www.globalcarbonproject.org/carbonbudget/19/infographics.htm.

13. The National Institute for Standards and Technology defines cyber-physical systems as comprising "interacting digital, analog, physical, and human components engineered for function through integrated physics and logic." "Cyber-Physical Systems," National Institute for Standards and Technology, accessed November 10, 2020, https://www.nist.gov/el/cyber-physical-systems.

CHAPTER 2

1. In fact, that is the essence of the global warming problem: humanity rapidly releasing all those hydrocarbons that were trapped below the Earth's surface and whose removal from the atmosphere helped create our current Goldilocks "not too hot and not too cold" temperate climate we enjoy today.

2. "Electricity," Switch Energy Alliance, accessed November 12, 2020, http://www.switchenergyproject.com/education/CurriculaPDFs/SwitchCurricula-Secondary-Electricity/SwitchCurricula-Secondary-ElectricityFactsheet.pdf.

3. The first steam turbine was invented in England in 1884. See "How Does a Steam Turbine Work?" Petrotech, accessed November 12, 2020, https://petrotechinc.com/how-does-a-steam-turbine-work/. Wind turbines function similarly, with rotors and stators.

4. One watt is equal to one joule of energy per second. A joule in turn is equal to the energy transferred to an object when the force of one newton acts on an object for a distance of one meter, and one newton is the force necessary to accelerate one kilogram of mass at the rate of 1 meter/second^2 in the direction of the applied force.

5. "Frequently Asked Questions," EIA, last updated October 9, 2020, https://www.eia.gov/tools/faqs/faq.php?id=97&t=3.

6. "Rolling Out Residential Demand Charges," Brattle Group, May 2015, http://files.brattle.com/files/5934_rolling_out_residential_demand_charges_hledik_euci.pdf.

7. "Power Blocks in Natural Gas-Fired Combined-Cycle Plants Are Getting Bigger," EIA, February 12, 2019, https://www.eia.gov/todayinenergy/detail.php?id=38312. Gas-fired plants have increased in size in recent years, from averaging about 500 MW in the first half of this decade to over 800 MW by 2018 as gas prices fell and efficiencies increased.

8. In 2016, 381 U.S. coal plants provided 249,000 MW of capacity.

9. "Nuclear Explained," EIA, last updated April 15, 2020, https://www.eia.gov/energyexplained/nuclear/us-nuclear-industry.php. At the end of 2019, there were 96 U.S. reactors fielding 98,000 MW of generating capacity.

10. As of December 31, 2018, there were about 9,719 power plants in the U.S. capable of generating at least 1 MW. "Frequently Asked Questions," EIA.

11. "Geothermal Explained," EIA, last updated March 25, 2020, https://www.eia.gov/energyexplained/geothermal/use-of-geothermal-energy.php.

12. "Hydropower Explained," EIA, last updated March 30, 2020, https://www.eia.gov/energyexplained/hydropower/.

13. "2018 Wind Technologies Market Report," U.S. Department of Energy, 2018, https://emp.lbl.gov/sites/default/files/wtmr_final_for_posting_8-9-19.pdf.

14. "Resource Adequacy: What Is It and Why Should You Care?" Gridworks, June 17, 2018, https://gridworks.org/2018/06/resource-adequacy-what-is-it-and-why-should-you-care/.

15. "ISO Requested Power Outages Following Stage 3 Emergency Declaration; System Now Being Restored," California ISO, August 15, 2020, http://www.caiso.com/Documents/ISORequestedPowerOutagesFollowingStage3EmergencyDeclarationSystemNowBeingRestored.pdf.

16. Markets FAQs," PJM Learning Center," accessed November 12, 2020, https://learn.pjm.com/three-priorities/buying-and-selling-energy/markets-faqs/what-are-black-start-services.aspx.

17. For example, grid operator ERCOT (Electric Reliability Council of Texas) does not have a market for capacity. Instead, they let prices soar to as high as $9,000 as an incentive for generators to come to the state and build power plants.

18. "Electricity," Switch Energy Alliance.

19. "United States Electricity Industry Primer," U.S. Department of Energy Office of Electricity Delivery and Energy Reliability, July 2015, http://www.emnrd.state.nm.us/ecmd/LinksNEW/documents/united-states-electricity-industry-primer.pdf.

20. Even though aluminum wires have to be 50% thicker to enable the same current as a copper wire, it's twice as light. "Aluminium in Power Engineering,"

All about Aluminum, accessed November 12, 2020, https://www.aluminium-leader.com/application/electrical_engineering/.

21. "Frequently Asked Questions," EIA.

22. Jordan Wirfs-Brock, "Lost in Transmission: How Much Electricity Disappears between a Power Plant and Your Plug?" Inside Energy, November 6, 2015, http://insideenergy.org/2015/11/06/lost-in-transmission-how-much-electricity-disappears-between-a-power-plant-and-your-plug/.

23. "How Power Grids Work," Clark Science Center, Smith College, accessed November 12, 2020, http://www.science.smith.edu/~jcardell/Courses/EGR220/ElecPwr_HSW.html.

24. A good example of the playing out occurred in Maine. A 1,200 MW power line proposed by Central Maine Power to bring hydroelectricity from Quebec to Massachusetts had been the subject of intense debate since it was proposed, and came to a head in the summer of 2020. A portion of the line would cut through about 50 miles of Maine woods. By August, Central Maine Power and Hydro-Quebec had spent $16.7 million in a campaign to support the project. For its part, the opposition was being supported by almost $6 million in television ads underwritten by two companies owning large gas-fired generating plants in Maine. See Steve Mistler, "Gas Companies to Spend $6 Million Encouraging Voters to Oppose Corridor," *Bangor Daily News*, July 16, 2020, https://bangordailynews.com/2020/07/16/politics/gas-companies-to-spend-6-million-encouraging-voters-to-oppose-corridor/.

25. Jeff St. John, "7 Transmission Projects That Could Unlock a Renewable Energy Bounty," Greentech Media, April 9, 2020, https://www.greentechmedia.com/articles/read/9-transmission-projects-laying-the-paths-for-cross-country-clean-energy.

26. "About," SOO Green HVDC Link, accessed November 12, 2020, https://www.soogreenrr.com/about/. The company launched an open solicitation process in August 2020, so the completion of this project appears increasingly likely.

27. Catherine Morehouse, "FERC Staff to Congress: HV Transmission Essential to Reducing Carbon, Deploying Renewables," Utility Dive, August 15, 2020, https://www.utilitydive.com/news/ferc-staff-to-congress-hv-transmission-essential-to-reducing-carbon-deplo/583177/?utm_source=Sailthru&utm_medium=email&utm_campaign=Newsletter%20Weekly%20Roundup:%20Utility%20Dive:%20Daily%20Dive%2008-15-2020&utm_term=Utility%20Dive%20Weekender.

CHAPTER 3

1. https://www.bloomberg.com/news/articles/2021-02-16/skyrocketing-texas-power-prices-may-enrich-some-bankrupt-others?sref=xfyiavTX.

2. https://www.bloomberg.com/news/articles/2021-02-22/texans-will-pay-for
-the-state-s-power-crisis-for-decades-to-come?sref=xfyiavTX.

3. https://www.eia.gov/electricity/monthly/epm_table_grapher.php?t=table
_5_02.

4. https://www.bloomberg.com/news/articles/2021-02-27/griddy-barred-from
-texas-power-market-for-payment-breach?sref=xfyiavTX.

5. Economists refer to changes in consumption relative to changes in price as
"elasticities," and they apply numerical values to measure that response function.
For example, if the price of a commodity doubles while demand falls by exactly
50%, the elasticity would be 1.0. In general, consumers of electricity do not exhibit
elastic behavior for several reasons: 1) they don't see the scarcity prices reflected
in their costs—or at least not in real time; 2) they may not have the technology
necessary to effect changes even if they wanted to; and 3) they may not be able
or willing to change consumption patterns. A classic example of inelastic behavior
would be an elderly couple using air conditioning during hot summer days. The
AC might simply be critical in maintaining their quality of life or even safety. So,
they might be able to dial the AC back a little, but have minimal flexibility in the
matter. McAlpin, by contrast, exhibits highly elastic behavior.

6. A smart meter is an electronic utility meter that records electricity consumption and frequently communicates the usage information back to the utility. In
some areas, it may also provide that information to other third parties if the consumer agrees to share that information. Smart meters provide much more detailed
information on electricity consumption within designated timeframes than the old
spinning disk analog meters, which were typically read only once per month. The
new meters often record consumption on a 5- or 15-minute, or an hourly, basis and
ship the information back to the utility at least once daily. They are now in place
for residential customers in more than half the country. See "Frequently Asked
Questions," EIA, accessed November 13, 2020, https://www.eia.gov/tools/faqs/
faq.php?id=108&t=3. Previously, the old familiar analog meters with the spinning
disks were read manually by meter readers once a month.

7. Chris Martin and Naureen Malik, "Texas Power Prices Briefly Surpass $9,000
Amid Scorching Heat," Bloomberg.com, August 12, 2019, https://www.bloom-
berg.com/news/articles/2019-08-12/searing-texas-heat-pushes-power-prices-to-
near-record-levels.

8. Mark Watson, "ERCOT Sets Peakload Record; Real-Time Prices Top
$6,000/MWh," S&P Global, August 12, 2019, https://www.spglobal.com/platts/
en/market-insights/latest-news/electric-power/081219-ercot-sets-peakload-re-
cord-real-time-prices-top-6000-mwh.

9. In 2019, an equity investment from French power giant EDF allowed it to
expand beyond Texas into additional markets in the northeastern United States.
Gerald Porter Jr. and Chris Martin, "Texas Startup Letting Homeowners Buy
Wholesale Power Gets Funding to Expand," Bloomberg.com, May 14, 2019,

https://www.bloomberg.com/news/articles/2019-05-14/edf-backs-texas-startup-letting-homeowners-buy-wholesale-power.

10. "About Us," Griddy, accessed November 13, 2020, https://www.gogriddy.com/about/.

11. This will be discussed in much greater detail in chapter 10 on batteries and energy storage.

12. Mark Watson, "ERCOT Expects Summer 2019 Planning Reserve Margin to Be Below Target," S&P Global, December 4, 2018, https://www.sp-global.com/marketintelligence/en/news-insights/trending/njHS_s048R23CAW-RQbfFGw2. That 13.75% number was based on engineering studies and set to ensure that blackouts, dryly termed "firm load shedding," had a chance of occurring no more than one day every decade. Recent analysis suggests that with a growing number of backup generators at the customer site and other assets—such as a growing population of batteries that are not visible to grid planners—that number can safely be lower.

13. The FERC does allow for other costs to be recouped by generators in extreme cases, as well as a number of additional charges that are not featured in the Texas market.

14. As distinguished from the price the utility charges for delivery across its natural monopoly of wires and poles.

15. In some other power grids, like those in the mid-Atlantic states and New England, generators are paid both for the energy—the megawatts per hour—they generate, as well as for their *ability* to generate energy at any specific time (called installed capacity). These payments may represent well over 20% of total revenue streams. In order to conceptualize this, it may help to use an imperfect analogy of a personal taxi just for you, the reader: imagine a taxi that waits outside your residence, waiting to drive you to your destination whenever you need transportation. In markets with capacity payments—such as New England and the mid-Atlantic power pool—the taxi driver would get paid a fee simply for being there, ready to drive you to your destination, irrespective of whether you take one ride per year or 100. So, you might pay that driver $50,000 a year just to be there and available. Under that model, in addition to that fixed fee, you would also pay something for each ride, based on the cost of gasoline, the driver's hourly wage, and depreciation costs of the vehicle. Maybe each trip would cost $100. All of those payments taken together would have to be enough to incentivize the driver to be there 24/7 and for the operating costs per trip. At a minimum, the driver would have certainty of a $50,000 revenue stream. In Texas, by contrast, the taxi would wait outside all year and receive nothing for being constantly on call. It might even be years before the driver takes you anywhere. So, when you do jump into the car, the driver must make up for all of his/her fixed and operating costs. One trip might therefore cost $25,000, and perhaps $50,000, since all of the costs can be repaid only if a trip is taken. The electricity market is akin to that in Texas. Some years, the power plants on the margin never get called, or are called on only infrequently to meet that peak

demand. That's when they have to make all their revenues. Investors must have some reasonable confidence that when they build a new plant, revenues will be sufficient to yield a profit. Oversimplified, that's how and why prices can soar 300 times the long-term average (spot market) real-time price.

16. Much of that flow of electricity is *not* converted into useful outcomes. For example, a typical incandescent lightbulb represents a worst-case scenario; roughly 95% of the electrons—moving through that tungsten filament of resistance until it glows—are wasted in the form of heat. More efficient LEDs (light emitting diodes) use only about 75% of that energy to achieve the same outcome.

17. In fact, in the summer that AC probably started in the early morning to pre-cool the building well before the occupants arrive.

18. This is especially true if the three or four days in a row are all weekdays, since weekday demands are significantly higher than that of holidays or weekends.

19. "Project No. 49852, Review of Summer 2019 ERCOT Market Performance," ERCOT, October 8, 2019, http://www.ercot.com/content/wcm/lists/172485/Review_of_ERCOT_Summer_2019_-_PUC_Workshop_-_FINAL_10-8-19.pdf. While August was hot, average summer temperatures were actually on the low side, at just the 21st hottest summer recorded in ERCOT.

20. ERCOT Prices Spiked to Historic Levels—And It Can Happen Again," Direct Energy, August 15, 2019, https://business.directenergy.com/blog/2019/august/ercot-prices-spiked-to-historical-levels.

21. "About Exelon: America's Leading Energy Provider," Exelon, accessed November 13, 2020, https://www.exeloncorp.com/company/about-exelon.

22. Since gas is both used for heating *and* for electricity generation, cold weather can increase gas demand for heating, which in turn pushes up the price of gas used in generating power, resulting in higher energy prices. The relationships can be quite complex, and not always constant.

23. Power flows 24/7, so the trading floor is always "on."

24. "About Supercomputers," National Weather Service, accessed November 13, 2020, https://www.weather.gov/about/supercomputers. NOAA vastly upgraded its computers in 2018, allowing the National Weather Service to implement its upgraded Global Forecast System (GFS) model. The combined capabilities allow NOAA to run the model with higher resolution and 16 days out, compared with 13 kilometers and 10 days in the prior model. See also "NOAA Kicks Off 2018 with Massive Supercomputer Upgrade," NOAA, January 10, 2018, https://www.noaa.gov/media-release/noaa-kicks-off-2018-with-massive-supercomputer-upgrade. Faster machines also allow forecasts to run better versions of "model ensembles"—those things frequently referred to on the Weather Channel, which are simply models run in multiple iterations after slightly different inputs are used to help assess the probability of various weather outcomes (see Andrew Freedman, "The Weather Service Is Getting Better Supercomputers," January 6, 2015, https://mashable.com/2015/01/06/weather-service-supercomputers-petaflops/). Faster and more powerful machines result in improved turnaround time for simula-

tion runs, so that weather events can be predicted ahead of time. Better computers also create higher fidelity, with improved ability to model weather in smaller grid sizes—reduced from 2 km^2 down to 1 km. That in turn creates more accurate local forecasts, allowing for better prediction of localized phenomena such as fog, thunderstorms, downpours, and hail.

25. Thousands of megawatts of new solar arrays will be brought online in the next few years, in part driven by soon-to-expire federal investment tax credits, and they will produce energy between roughly 9:00 a.m. and 3:00 p.m. Thus, cheap renewables will cover more hours of the day, making it more difficult for other generators to make a profit.

26. "Capacity and Demand Reserves Report," ERCOT, December 2019, http://www.ercot.com/gridinfo/resource. The capacity number represents available capacity that could theoretically meet demand if necessary. A gas-fired generator is fully dispatchable (except when down for outages and maintenance), so it would get a full capacity value. Renewable resources are intermittent and would not be accorded the same value. They don't always generate electrons when one wants and needs them to do so.

27. "Capacity Changes by Fuel Type Charts," ERCOT, December 2019, http://www.ercot.com/gridinfo/resource. The majority of these resources already have an interconnection agreement with ERCOT and have posted financial security. It should be noted that for planning purposes, the wind resources do not get accorded a full capacity number. Coastal wind, which blows during daytime hours and generally coincides with peak demand, is accorded a 63% peak average capacity contribution (63% of installed capacity). By contrast, Panhandle wind is a nighttime phenomenon, so it gets a far lower value of 29%, while utility-scale solar gets a value of 76%.

CHAPTER 4

1. "Electricity," Merriam-Webster.com Dictionary, accessed December 1, 2020, https://www.merriam-webster.com/dictionary/electricity.

2. Seven ISOs/RTOs oversee generation and transmission assets in competitive and wholesale markets covering roughly half the geographic territory of the United States, and accounting for approximately 60% of the electricity that gets generated and consumed; see "About 60% of the U.S. Electric Power Supply Is Managed by RTOs," EIA, April 4, 2011, https://www.eia.gov/todayinenergy/detail.php?id=790. In the southeastern and much of the western parts of the country where there are no competitive wholesale markets, the vertically integrated utilities continue to manage the entire system. The competitive markets with wholesale trading include California—which also includes part of Nevada; the Southwest Power Pool, stretching from a portion of Idaho and North Dakota as far south as the northern portion of Texas and eastern New Mexico; most of the remainder

of Texas; the Mid-Continental system, reaching from other patchwork portions of Idaho and North Dakota all the way to a portion of Mississippi, Louisiana, and portion of eastern Texas; and the mid-Atlantic region encompassing areas from as far west as Indiana east to a portion of Virginia, and north to Pennsylvania and New Jersey. Finally, in the northeast, New England and New York each constitute separate markets.

3. Dispatchable means that these power plants can respond within prescribed timeframes to a signal from the ISO, either starting up or shutting down in order to maintain that constant critical balance between supply and demand.

4. The exception to that are nuclear plants, and to some degree coal plants as well. Nuclear plants are difficult to dial back, simply because of the nature of the fission process. Coal plants also do not function that well when made to operate in a variable fashion and "ramped" up or down frequently. It increases the wear and tear on the facilities, with their massive boilers and turbines.

5. Anna Duquiatan, Taylor Kuykendall, Darren Sweeney, and Liz Thomas, "US Power Generators Set for Another Big Year in Coal Plant Closures in 2020," S&P Global, January 13, 2020, https://www.spglobal.com/marketintelligence/en/news-insights/latest-news-headlines/56496107. This phenomenon is occurring across the United States. One estimate has the percentage of coal-fired generation declining from 27% of the U.S. generation mix in 2018 to as little as 8% by 2030.

6. "Generator Interconnection Queue," ISO–New England IRTT System, updated daily, https://irtt.iso-ne.com/reports/external. An examination of the queue shows hundreds of smaller resources, many of them solar projects and batteries, lining up to connect to the grid.

7. In the current queues, especially on the East Coast, there are a small number of enormous offshore wind assets in the 800–1,000 MW range, followed by a multitude of much more modest resources—often solar and battery storage projects as small as 10 or 20 MWs. Only a small percentage of assets in the interconnection queue eventually get connected to the grid and generate electricity, for a number of reasons. Many assets are entered into the queue by developers simply so that they have an option to connect at a later date, assuming they can line up permits, land, equipment, and most important of all—financing. In addition, many assets may lose viability if other competing assets are brought to market ahead of them. So, if asset A gets financed and approved, it renders asset B nonviable.

8. In the presentation prepared for my visit to ISO-NE, the proposed wind resource was listed as 14,256 MW. Broken down, Maine accounted for 751 MW of that target, tiny Rhode Island came in at 880 MW, Connecticut called for 4,160 MW, and Massachusetts made up well over half at 8,460 MW. The initial two 800+ MW projects have already been awarded. The first—Vineyard Wind—had an expected commissioning date around 2022/2023, but this has been pushed back owing to delays in the federal permitting process. Project-level information is available at ISO-New England IRTT System, https://irtt.iso-ne.com/reports/external.

9. Ibid.

10. Many of these resources are highly subsidized in states like Massachusetts, and paid for by an energy-efficiency charge on the customer's utility distribution bill. For example, in February 2019, my local hardware store had regular LED bulbs and LED spotlights for sale at 99 cents. The same bulb was for sale online at $12.31. "MAXLITE 14099210," Regency Lighting, accessed November 14, 2020, https://shop.regencylighting.com/15p38wd27fl-mxt.html?gclid=EAIaIQobChMI9dvE7bDW5wIVAxgMCh3z7gvYEAQYASABEgLxavD_BwE.

11. These losses can total as much as 6%–8% due to the heat created when moving a steady flow of electrons through power lines. Such losses increase in the summer when the lines are warmer.

12. An investigation into the deadly 2018 Camp Fire that killed 85 people and incinerated close to 19,000 homes laid the blame for the fatal blaze on northern California utility Pacific Gas & Electric's transmission lines: Peter Eavis and Ivan Penn, "California Says PG&E Power Lines Caused Camp Fire That Killed 85," *New York Times*, May 15, 2019, https://www.nytimes.com/2019/05/15/business/pge-fire.html. Subsequently, the utility has preemptively shut power to tens of thousands of customers during high wind events, sometimes for days.

13. "California Household Battery Sales to Quadruple in 2020," Bloomberg NEF, February 10, 2020, https://about.bnef.com/blog/california-household-battery-sales-to-quadruple-in-2020/.

14. Latest numbers from December 2018 were made available at the time of my meeting; see "Final 2019 PV Forecast," ISO-NE, April 29, 2019, https://www.iso-ne.com/static-assets/documents/2019/04/final-2019-pv-forecast.pdf. This number had increased by 500 MW and 24,600 installations from the previous year's forecast, available at https://www.iso-ne.com/static-assets/documents/2018/04/final-2018-pv-forecast.pdf.

15. "Cumulative Installed PV as of Dec 31, 2019," Hawaiian Electric, accessed November 14, 2020, https://www.hawaiianelectric.com/documents/clean_energy_hawaii/clean_energy_facts/pv_summary_4Q_2019.pdf; "Power Facts," Hawaiian Electric, last updated December 31, 2019, https://www.hawaiianelectric.com/about-us/power-facts.

16. Trieu Mai, Paige Jadun, Jeffrey Logan, Colin McMillan, Matteo Muratori, Daniel Steinberg, and Laura Vimmerstedt, "Electrification Futures Study: Scenarios of Electric Technology Adoption and Power Consumption for the United States," National Renewable Energy Laboratory, 2018, https://www.nrel.gov/docs/fy18osti/71500.pdf.

17. For New England, that means having a "reserve margin" or "reserve requirements" of sufficient extra generating capability equal to the sum of 100% of first largest contingency (generating asset or transmission line) plus 50% of second largest asset. The amount of assets or "reserve" required to replace the largest contingency must be capable of responding within 10 minutes, with about 25% of that resource synchronized to the actual frequency of the AC grid. The reserves required to replace 50% of the second largest contingency must be capable of

responding within 30 minutes. So, if New England were to lose its largest re-source—in this case a transmission line in from Quebec—it could quickly access sufficient energy from other resources to keep the grid stable. See "2018 Annual Markets Report," ISO New England, May 23, 2019, https://www.iso-ne.com/static-assets/documents/2019/05/2018-annual-markets-report.pdf. In recent years, total reserve requirements have totaled about 2,500 MW. According to ISO-NE, the total installed capacity requirement is "the amount of capacity . . . needed to meet the region's reliability requirements (including energy and reserves). The ICR requirements are designed such that non-interruptible customers can expect to have their load curtailed not more than once every ten years." Other regions often prefer a fixed percentage target, though they may not always meet it. In 2019, the Electric Reliability Council of Texas (ERCOT) carried an 8.6% reserve margin, in comparison with the stated 13.75% target ERCOT had set for itself (see "Resource Adequacy," ERCOT, accessed November 14, 2020, http://www.ercot.com/gridinfo/resource). That number increased to 10.6% in 2020. For its part, PJM—the mid-Atlantic grid operator—currently recommends a reserve margin hovering close to 15% for the foreseeable future ("2019 PJM Reserve Requirement Study," PJM, October 8, 2019, https://www.pjm.com/-/media/committees-groups/sub-committees/raas/20191008/20191008-pjm-reserve-requirement-study-draft-2019.ashx). Those assets must be paid for by energy consumers, so higher reserve margins are often a source of concern for those who foot the bill.

18. In just the single month of January 2020, a senior Connecticut state environmental official made noises about leaving the ISO, citing a market "that is driving investment in more natural gas and fossil fuel power plants that we don't want and we don't need. . . . This is forcing us to take a serious look at the cost and benefits of participating in the ISO New England markets" (Patricia Skahill, "CT Taking 'A Serious Look' at Exiting Regional Power Market," CT Mirror, January 16, 2020, https://ctmirror.org/2020/01/16/conn-taking-a-serious-look-at-exiting-regional-power-market/). Meanwhile, in the same month, the governor of Illinois asked the state legislature to pass clean energy laws that may result in the state leaving the mid-Atlantic (PJM) grid. The policies requiring grid operators to treat subsidized resources differently have emanated from the short-staffed FERC, which was dominated by two Republican commissioners in 2020. The lone Democrat issued stinging dissents on more than one occasion; see "Illinois Governor Wants Clean Energy Legislation, Could Push State Out of PJM Power Grid," Reuters, January 30, 2020, https://www.reuters.com/article/us-usa-illinois-pjm-nuclearpower/illinois-governor-wants-clean-energy-legislation-could-push-state-out-of-pjm-power-grid-idUSKBN1ZT259. At stake—among other things—are the current subsidies the legislature voted for to support the existing, economically struggling nuclear power plants. PJM has changed the rules affecting the bidding process into the market, effectively putting at risk or neutralizing the subsidies. In early 2020, the FERC effectively mandated a similar approach to New York State's wholesale market, with an expected negative impact on some of the most aggressive carbon

policies in the country. Some observers believe this issue may eventually find its way to the Supreme Court.

19. Susan F. Tierney and Paul J. Hibbard, "Clean Energy in New York State: The Role of and Economic Impacts of a Carbon Price in NYISO's Wholesale Electricity Markets," NYISO, October 3, 2019, https://www.nyiso.com/documents/20142/2244202/Analysis-Group-NYISO-Carbon-Pricing-Final-Summary-for-Policymakers.pdf/75a766a8-623f-c105-ddcf-43dd78cb4bca?t=1570098881971.

20. The conversation was both for this book and for an article I was writing at that time (Peter Kelly-Detwiler, "New York Power Grid Proposes Adding Carbon Costs To Market Prices For Electricity," Forbes.com, February 20, 2020, https://www.forbes.com/sites/peterdetwiler/2020/02/20/in-a-path-breaking-approach-new-yorks-grid-operator-proposes-inclusion-of-carbon-costs-in-market-prices/#5cb0f9b650af).

21. In fact, the FERC held a hearing on carbon pricing at the end of September 2020, to better understand the implications of this approach.

22. Robert Walton, "ISO New England Chief Presses for Carbon Price in Response to Sanders, Warren and Others," Utility Dive, November 25, 2109, https://www.utilitydive.com/news/iso-new-england-chief-presses-for-carbon-price-in-response-to-sanders-warr/567955/.

23. The U.S. grid transmits energy in alternating current (AC) mode. Unlike direct current (DC), where the flow simply moves in one direction, AC involves oscillations of the flow. In the United States this happens 60 times per second, expressed as 60 Hz. The power coming from our wall sockets is AC, but many devices, such as computers, require DC. In those instances, the flow of AC energy must be converted to DC by a transformer. That little box that powers your laptop does just that.

24. "Success Factors in System Operator Selection," PSP Metrics, accessed November 14, 2020, http://www.psptesting.com/articles/Success%20Factors%20in%20System%20Operator%20Selection%20abbr.pdf. PSP comments on its website that it is difficult to identify successful system operators for three reasons: "First, situations do not repeat frequently in the control room setting. Second, it is impossible to predict all of the problems that can arise on the job. Third, usually a relatively small number of system operators work in any one control center." Further, there is a lengthy learning curve requiring extensive and costly training. "Thus, it takes a great deal of time before a manager knows for certain if an operator will succeed," with that timeframe often over a year.

25. http://www.ercot.com/content/wcm/key_documents_lists/225373/2.2_ERCOT_Presentation.pdf.

26. https://www.bloomberg.com/news/features/2021-02-20/texas-blackout-how-the-electrical-grid-failed?sref=xfyiavTX.

27. https://www.statesman.com/story/news/2021/02/25/texas-power-outage-death-toll-medical-examiner-processing-86-cases/6808150002/.

28. Hannah Northey and David Iaconangelo, "'We Have Never Done This Before': Inside N.Y.'s Grid Lockdown," E&E News, March 30, 2020, https://www.eenews.net/stories/1062737239. This unprecedented decision required the operators to live at facilities, 24/7. To ensure the state's power keeps flowing, NYISO also has backup plans in the event that both control rooms were to become simultaneously infected, with local utilities Consolidated Edison and National Grid PLC to help operate the grid in a worst-case scenario. CEO Richard Dewey was quoted as saying, "We have never done this before. We drilled for this stuff, we've had plans in place for different types of sequestration. This is pretty unprecedented in our history."

29. "April and May 2018 Fault Induced Solar Photovoltaic Resource Interruption Disturbance Report," North American Electric Reliability Corporation (NERC), January 2019, https://www.nerc.com/pa/rrm/ea/April_May_2018_Fault_Induced_Solar_PV_Resource_Int/April_May_2018_Solar_PV_Disturbance_Report.pdf. In this case, both utility-scale and distributed solar arrays went offline. In the 2016 California Blue Cut fire, 1,200 MW of utility-scale solar projects went temporarily offline (see "1,200 MW Fault Induced Solar Photovoltaic Resource Interruption Disturbance Report," NERC, June 2017, https://www.nerc.com/pa/rrm/ea/1200_MW_Fault_Induced_Solar_Photovoltaic_Resource_/1200_MW_Fault_Induced_Solar_Photovoltaic_Resource_Interruption_Final.pdf). In investigating that incident, the North American Reliability Corporation, whose job it is to ensure reliable and secure supplies of electricity, noted that the local utility, Southern California Edison (SCE), and the California Independent System Operator (CAISO) were surprised by what they did not know. "Now aware of the potential for this action, SCE/CAISO discovered that this was not an isolated incident. Including the August 16 events, SCE/CAISO determined that this type of inverter disconnect has occurred eleven times between August 16, 2016, and February 6, 2017."

CHAPTER 5

1. "Cori," National Energy Research Scientific Computing Center (NERSC), accessed November 15, 2020, https://docs.nersc.gov/systems/cori/.

2. The race for larger and more powerful machines is a global one, with two major competitors: the Americans and the Chinese. Today, the United States sits in the number one and two spots, with Oak Ridge National Laboratory's Summit (148.6 petaflops) in pole position and Lawrence Livermore National Laboratory's Sierra in second place at 94.6 petaflops. Not too far behind is China's Sunway TaihuLight supercomputer, at 93.0 petaflops. Cori currently sits in 13th place. See "China Extends Lead in Number of TOP500 Supercomputers, US Holds on to Performance Advantage," Top 500, 2020, https://www.top500.org/news/china-extends-lead-in-number-of-top500-supercomputers-us-holds-on-to-performance-advantage/.

3. Gerbrand Ceder and Kristin Persson, "How Supercomputers Will Yield a Golden Age of Materials Science," Scientific American, December 1, 2013, http://www.scientificamerican.com/article/how-supercomputers-will-yield-a-golden-age-of-materials-science/.

4. Although Gerty Cori, who was awarded the Nobel Prize in Physiology or Medicine in 1947, passed away 10 years later, Nobel Laureate Saul Perlmutter, for whom the next supercomputer is named, is still very much alive. Perlmutter was awarded the 2011 Nobel Prize in Physics "for the discovery of the accelerating expansion of the Universe through observations of distant supernovae" while working at none other than Lawrence Berkeley National Laboratory.

5. There are larger computers at other national laboratories, including Oak Ridge National Laboratory's Summit (148.6 petaflops), with "exa-scale" computers (exceeding 1,000 petaflops) being developed for both Oak Ridge and Argonne National Laboratories. However, these are not designed for the types of materials science inquiries Simon and Persson oversee. Persson indicated that NERSC is better suited for the basic fundamental research. NERSC has more users, in part since more codes can be run on NERSC's less specialized systems. Code development, explained Persson, takes longer than developing new architectures. As a consequence, the more powerful machines such as Summit would not exploit the computational power very effectively. "When I tried to run at Oak Ridge Titan," said Persson, "I had to run at 25% of capability just to keep memory available for the codes I am using."

6. The higher density chemistry means it packs more punch for the same size. Persson keeps one on her desk at LBL, a testament to the potential of the investigative process. "Upgrade to the Power Of Duracell Optimum," Duracell, accessed November 15, 2020, https://www.duracell.com/en-us/product/optimum-battery/.

7. Katharine Sanderson, "Automation: Chemistry Shoots for the Moon," Nature, 568, April 23, 2019, https://www.nature.com/articles/d41586-019-01246-y.

8. "Frequently Asked Questions," EIA, last updated November 18, 2020, https://www.eia.gov/tools/faqs/faq.cfm?id=65&t=3. These power plants are over 1 MW in size.

9. "Transmission," Edison Electric Institute (EII), accessed November 15, 2020, http://www.eei.org/issuesandpolicy/transmission/Pages/default.aspx. This makes the grid over four times the length of the U.S. federal highway system.

10. Byron Boyle, "What a Nodal Market Means for Texas and ERCOT," Webb Energy Blog, March 14, 2010, https://webberenergyblog.wordpress.com/2010/03/14/what-a-nodal-market-means-for-texas-and-ercot/.

11. Khrosrow Moslehi, "Texas Transformed," T&D World, December 9, 2013, https://www.tdworld.com/grid-innovations/article/20963765/texas-transformed.

12. "History" (video), ERCOT, accessed November 15, 2020, http://www.ercot.com/about/profile/history.

13. "Nodal Program Revised Preliminary Budget," Special Board of Director's Meeting, ERCOT, January 21, 2009, http://www.ercot.com/content/meetings/board/keydocs/2009/0121/Nodal_Program_Revised_Preliminary_Budget.pdf.

14. Jim Lazar and Xavier Baldwin, "Valuing the Contribution of Energy Efficiency to Avoided Marginal Line Losses and Reserve Requirements," Regulatory Assistance Project, August 2011, http://www.raponline.org/wp-content/uploads/2016/05/rap-lazar-eeandlinelosses-2011-08-17.pdf.

15. Jan Ellen Spiegel, "Another $1.2 Billion Substation? No Thanks, Says Utility, We'll Find a Better Way," Inside Climate News, April 4, 2016, https://insideclimatenews.org/news/04042016/coned-brooklyn-queens-energy-demand-management-project-solar-fuel-cells-climate-change.

16. For all things REV, this link is a good place to start: "About the Initiative," New York State DPS: Reforming the Energy Vision," Last Updated May 9, 2018, http://www3.dps.ny.gov/W/PSCWeb.nsf/All/CC4F2EFA3A23551585257DEA007DCFE2?OpenDocument.

17. David Labrador, "New York REV's Distributed System Platform Breaks New Ground," Rocky Mountain Institute, August 27, 2015, https://rmi.org/blog_2015_08_27_new_york_rev_distributed_platform_breaks_new_ground/.

18. "FERC Order No. 2222: A New Day for Distributed Energy Resources" (fact sheet), Federal Energy Regulatory Commission, September 17, 2020, https://www.ferc.gov/sites/default/files/2020-09/E-1-facts.pdf.

19. "Power Generation: Moving into an Era of Digital Energy," GE Digital, accessed November 15, 2020, https://www.ge.com/digital/sites/default/files/Power%20Digital%20Solutions%20Product%20Catalog.pdf.

20. There is, in fact, an Industrial Internet of Things Consortium, comprised of over 200 members. A number of these are large power companies such as ABB, GE, Mitsubishi, and Toshiba, to name a few. The consortium focuses on a disparate range of issues. Chief among them are standards, best practices, and cybersecurity. See Industrial Internet Consortium, http://www.iiconsortium.org.

21. "Power Generation: Moving into an Era of Digital Energy," GE Digital.

22. Companies such as GE with its Predix platform are working to build these capabilities. In many cases, they are opening their systems up to the broader development and consulting communities. GE, for example, lists numerous global partners ranging from Microsoft to Verizon (see "Our Ecosystem," GE Digital, accessed November 15, 2020, https://www.ge.com/digital/content/meet-our-ecosystem). Other manufacturers such as Mitsubishi and Siemens are adopting similar approaches.

CHAPTER 6

1. A PPA is a financial structure in which the owner of the home pays a specific amount for every kilowatt-hour generated. These contracts often have escalators that increase over time.

2. Not everybody can, and third-party finance has been absolutely critical to the growth of an industry that requires plunking down $20,000 or more. In states such

as Arizona, Colorado, and California, third-party financing has supported more than 80% of residential solar installations. See "Third-Party Financing," Center for the New Energy Economy, 2016, https://spotforcleanenergy.org/wp-content/uploads/2016/05/ff5c108c070020d3ed779350e45e2b77.pdf.

3. Tom Randall, "No One Saw Tesla's Solar Roof Coming," Bloomberg, October 31, 2016, https://www.bloomberg.com/news/articles/2016-10-31/no-one-saw-tesla-s-solar-roof-coming?sref=tSsSu99y. Randall commented, "Like everyone else, I knew we were there to see Musk's new "solar roof," whatever that was supposed to mean. But try as I could as we walked in, I didn't see anything that looked like it could carry an electric current. If anything, the slate and Spanish clay roofs looked a bit too nice for a television set."

4. Julian Spector, "5 Reasons Not to Get Too Excited About Tesla's New Solar Roof," GTM, October 5, 2019, https://www.greentechmedia.com/articles/read/5-reasons-to-not-get-too-excited-about-teslas-new-solar-roof.

5. Ibid.

6. I've profiled EnergySage in Forbes.com before, and have spoken with customers. The website (https://www.energysage.com/) is a valuable resource in many ways. EnergySage gives one the ability to compare quotes and also has equipment ratings (with numerous panels included), and easy-to-understand explanations for many various aspects of the industry.

7. For residential solar, the ITC is a credit you can apply against tax liabilities. It started at 30%, stepped down to 26% in 2020, drops to 22% in 2021, and goes to zero thereafter. For utility-scale and commercial installations, it follows the same trajectory, but pegs out at 10% in 2022 and thereafter, with no expiry date at present.

8. Forty states have (or have had) some sort of community solar program, but only a few states have sizable commitments to this approach. These include Colorado, Illinois, Massachusetts, Minnesota, and New York.

9. I am locked in for 20 years, but can transfer to another ratepayer within my utility service territory if the need ever arises.

10. "United States Surpasses 2 Million Solar Installations," Solar Energy Industries Association (SEIA), May 9, 2019, https://www.seia.org/news/united-states-surpasses-2-million-solar-installations.

11. Ibid.

12. Galen Barbose and Naïm Darghouth, "Tracking the Sun: Pricing and Design Trends for Distributed Photovoltaic Systems in the United States," Lawrence Berkeley National Laboratory, October 2019, https://emp.lbl.gov/sites/default/files/tracking_the_sun_2019_slide_deck_summary_0.pdf.

13. "Solar Industry Research Data," Solar Energy Industries Association, accessed November 16, 2020, https://www.seia.org/solar-industry-research-data.

14. "How to Reach and Acquire Residential Solar Customers," PV Solar Report, October 29, 2014, http://pvsolarreport.com/how-to-acquire-residential-solar-customers/. In that instance, the timeframe was reduced by two-thirds.

15. Julian Spector, "SunRun Sees Online Solar Sales Bump After Initial Lockdown Slump," GTM, May 6, 2020, https://www.greentechmedia.com/articles/read/sunrun-earnings-coronavirus-online-sales.

16. "SunRun (RUN) Q1 2020 Earnings Call Transcript," Motley Fool, May 6, 2020, https://www.fool.com/earnings/call-transcripts/2020/05/07/sunrun-run-q1-2020-earnings-call-transcript.aspx. During the May 6, 2020, quarterly earnings call, CEO Lynne Jurich commented, "that's not going to be our normal way of operating, but it's, again, what's possible when you can start to streamline the whole system."

17. "Vermont Regulators Approve Pioneering GMP Energy Storage Programs to Save Customers Money and Increase Reliability," Green Mountain Power, May 21, 2020, https://greenmountainpower.com/news/vermont-regulators-approve-pioneering-gmp-energy-storage-programs/.

18. Alan Jones, "Small Batteries but High Stakes for New Hampshire Home Energy Storage Pilot," Energy News Network, July 27, 2020, https://energynews.us/2020/07/27/northeast/small-batteries-but-high-stakes-for-new-hampshire-home-energy-storage-pilot/.

19. "Vermont Regulators Approve Pioneering GMP Energy Storage Programs."

20. "ISO New England Awards SunRun Landmark Wholesale Capacity Contract," GlobeNewswire, February 7, 2019, https://www.globenewswire.com/news-release/2019/02/07/1712238/0/en/ISO-New-England-Awards-Sunrun-Landmark-Wholesale-Capacity-Contract.html.

21. Jeff St. John, "SunRun Lands Contract for 20MW Backup Battery-Solar Project in Blackout-Prone California," GTM, July 30, 2020, https://www.greentechmedia.com/articles/read/sunrun-lands-20mw-backup-battery-solar-contract-for-northern-california-communities.

22. William Driscoll, "SunPower to Sell 11 MW of Capacity in New England's Grid," PV magazine, March 17, 2020, https://pv-magazine-usa.com/2020/03/17/sunpower-to-sell-11-mw-of-capacity-in-new-englands-grid/.

23. Robert Walton, "Rocky Mountain Power to Operate Largest US Residential Battery Demand Response Project," Utility Dive, August 27, 2019, https://www.utilitydive.com/news/rocky-mountain-power-prepares-to-operate-largest-us-residential-battery-dem/561553/.

24. Peter Kelly-Detwiler, "In Australia with Solar Plus Storage, You Can Now Get All Your Electricity for One Low Fixed Monthly Price," Forbes.com, September 23, 2019, https://www.forbes.com/sites/peterdetwiler/2019/09/23/in-australia-with-solar-plus-storage-you-can-now-get-all-your-electricity-for-one-low-fixed-monthly-price/#14dafc403722.

25. "SunRun (RUN) Q1 2020 Earnings Call Transcript," Motley Fool.

26. "A Million-Mile Battery from China Could Power Your Electric Car," Bloomberg News, June 7, 2020, https://www.bloomberg.com/news/articles/2020-06-07/a-million-mile-battery-from-china-could-power-your-electric-car?sref=xfyiavTX.

27. "Brightbox," SunRun, accessed November 16, 2020, https://www.sunrun.com/solar-battery-storage/lg-chem-battery-specs.

28. "How Does Solar Plus Battery Backup Work?" Sunbridge Solar, accessed November 16, 2020, https://sunbridgesolar.com/blog/generac-pwrcell-home-battery-review.

29. "Powerwall," Tesla.com, accessed November 16, 2020, https://www.tesla.com/sites/default/files/pdfs/powerwall/Powerwall%202_AC_Datasheet_en_northamerica.pdf.

30. 2020 IONIQ Electric, Hyundai USA, accessed November 16, 2020, https://www.hyundaiusa.com/us/en/vehicles/ioniq-electric/compare-specs.

31. "Nissan to Create Electric Vehicle 'Ecosystem,'" Nissan Motor Corporation Newsroom, November 28, 2020, https://usa.nissannews.com/en-US/releases/release-81d8e341e3784b4fbcce44225f91cbaa-nissan-to-create-electric-vehicle-ecosystem.

32. Bengt Halvorson, "Nissan Leaf as Home Energy Device: Wallbox Will Soon Enable It in the U.S.," Green Car Reports, March 26, 2020, https://www.greencarreports.com/news/1127590_nissan-leaf-as-home-energy-device-wallbox-will-soon-enable-it-in-the-u-s.

33. Blagojce Krivevski, "Nissan Accepting Electricity as Payment for Parking," Electric Cars Report, July 31, 2020, https://electriccarsreport.com/2020/07/nissan-accepting-electricity-as-payment-for-parking/.

CHAPTER 7

1. I also have a relationship to another 20 panels about 70 miles to my west in a community solar array in Uxbridge, Massachusetts; these offset the rest of my electric bill and I get a 15% reduction on my bill as a consequence.

2. I did arrive home with 16 miles left on the battery and a "low EV battery" warning that I was down to 13%.

3. The terms module and panel are used interchangeably in the industry, and so are used in the same fashion here as well.

4. U.S. Solar Market Insight," Solar Energy Industries Association (SEIA), September 10, 2020, https://www.seia.org/us-solar-market-insight.

5. That's the case with most solar technologies used in the United States, though there is also a competitor technology known as thin film that is manufactured in an entirely different process.

6. The wafers are sliced from solar ingots—increasingly with diamond-wire saws to reduce the amount of loss (called kerf—think of sawdust) and then polished.

7. For example, SunPower planned on spinning out its Maxeon Solar Technologies manufacturing arm by the third quarter of 2020, to create two separate companies. See "SunPower Provides Update on Planned Spin-Off of Maxeon Solar Technologies," Cision PR Newswire, June 16, 2020, https://www.prnewswire.

com/news-releases/sunpower-provides-update-on-planned-spin-off-of-maxeon-solar-technologies-301077527.html.

8. And there have been many other efficiency gains in the balance of system costs, from the inverters that connect the DC panels to the AC grid, to lower panel racking costs, to improved permitting, interconnection, and inspection—which is increasingly going digital in the COVID-19 era.

9. "The Shockley Queisser Efficiency Limit," Solar Cell Central, accessed November 16, 2020, http://solarcellcentral.com/limits_page.html. Originally calculated in 1961, the Shockley Queisser Limit is a highly arcane name for a rock band that is literally begging to be adopted.

10. "What Are the Most Efficient Solar Panels on the Market? Solar Panel Cell Efficiency Explained," EnergySage, accessed November 16, 2020, https://news.energysage.com/what-are-the-most-efficient-solar-panels-on-the-market/.

11. "Six-Junction Solar Cell Sets Two World Records for Efficiency," Science Daily, April 14, 2020, https://www.sciencedaily.com/releases/2020/04/200414173255.htm. This conversion efficiency was achieved by testing with an exposure of light equal to the output of one sun. By concentrating the light to an equivalent of 143 suns, researchers were able to boost the efficiencies to 47.1%. The device contains 140 total layers of various materials, but is three times narrower than a human hair.

12. "About JinkoSolar Holding Co., Ltd," Jinko Solar, accessed November 16, 2020, https://www.jinkosolar.com/en/site/aboutus.

13. Sandra Enkhardt, "Global PV Capacity Additions Hit 115 GW in 2019, says IEA," PV Magazine, May 1, 2020, https://www.pv-magazine.com/2020/05/01/global-pv-capacity-additions-hit-115-gw-in-2019-says-iea/. China represented the largest solar market in 2019 with 30.1 GW, while the United States installed 13.3 GW. In 2019, seven of the top ten globule module manufacturers were from China. Only one—thin film manufacturer FirstSolar—was from the United States (see Michael Bloch, "Top 10 Solar Panel Manufacturers in 2019," Solarquotes, February 13, 2020, https://www.solarquotes.com.au/blog/solar-shipments-rankings-mb1405/).

14. "Lazard's Levelized Cost of Energy Analysis: Version 13.0," Lazard, November 2019, https://www.lazard.com/media/451086/lazards-levelized-cost-of-energy-version-130-vf.pdf.

15. "Most Utility-Scale Fixed-Tilt Solar Photovoltaic Systems Are Tilted 20 Degrees–30 Degrees," IEA, October 26, 2018, https://www.eia.gov/todayinenergy/detail.php?id=37372. The further north the panels are located, the more it makes sense to tilt them to receive direct sunlight.

16. Cabe Atwell, "The Biggest Utility-Scale Solar Farms on Earth," ElectronicDesign, March 14, 2019, https://www.electronicdesign.com/power-management/article/21807720/the-biggest-utilityscale-solar-farms-on-earth.

17. Eric Wesoff, "Solar Star, Largest PV Power Plant in the World, Now Operational," GTM, June 26, 2015, https://www.greentechmedia.com/articles/read/solar-star-largest-pv-power-plant-in-the-world-now-operational.

18. Mark Bolinger, Joachim Seel, and Dana Robson, "Empirical Trends in Project Technology, Cost, Performance, and PPA Pricing in the United States," Lawrence Berkeley National Laboratory, December 2019, https://emp.lbl.gov/sites/default/files/lbnl_utility-scale_solar_2019_edition_slides_final.pdf. For the most part, fixed-tilt systems that do not utilize trackers are either located in areas with less sun (such as the northeastern United States) or windier areas where installations are more susceptible to damage.

19. There are also dual-axis trackers that allow one to follow the sun on both north-south and east-west axes, but they typically increase the output by only another 5%–10%, so the added expenses have not been justified in most cases (see Jacob Marsh, "Solar Trackers: Everything You Need To Know," EnergySage, January 19, 2018, https://news.energysage.com/solar-trackers-everything-need-know/).

20. Alex Roedel, "Driving the Standard for Solar Tracker Wind Design," Nextracker, December 3, 2019, https://www.nextracker.com/2019/12/driving-the-standard-for-solar-tracker-wind-design/.

21. That platform, called Virbela, allowed me to "walk" to the podium, stand next to a conference organizer who introduced me to the audience, present my slides and speak to the attendees, and see avatars in the audience with hands raised for questions at the end of my presentation.

22. "Case Study: Replacing and Recycling 13,000 Solar Panels Damaged by West Texas Hailstorm," World Wind & Solar, July 3, 2019, https://worldwind-solar.com/case-study-replacing-recycling-13000-solar-panels-damaged-by-west-texas-hailstorm/. In that event, 13,000 panels were destroyed and had to be replaced.

23. Xiaojing Sun, "WoodMac: Bifacial Solar Market Set to Grow Tenfold by 2024," GTM, September 24, 2019, https://www.greentechmedia.com/articles/read/bifacial-solar-market-set-to-grow-tenfold-by-2024.

24. "Webinar: Understanding Bifacial's True Potential," Photovoltaik-Institut Berlin AG, October 7, 2019, https://pdfs.semanticscholar.org/b80f/d6c5f-c6701a6fdd91390e334e1075b63442d.pdf.

25. Ibid.

26. "First Quarter 2020 Earnings Conference Call," NextEra Energy, April 2020, http://www.investor.nexteraenergy.com/~/media/Files/N/NEE-IR/reports-and-fillings/quarterly-earnings/2020/Q1/1Q%202020%20Script_vF.pdf. It's not clear whether that was Rhode Island's use for four hours at average demand or during peak demand. Either way, even if it's a small state, that's a lot of energy.

27. "Why Choose 8minute?" 8minute Solar Energy, accessed November 16, 2020, https://www.8minute.com.

28. Julian Spector, "US Storage Industry Achieved Biggest-Ever Quarter and Year in 2019," GTM, March 10, 2020, https://www.greentechmedia.com/articles/read/us-storage-industry-achieved-biggest-ever-quarter-year-in-2019?utm_medium=email&utm_source=Daily&utm_campaign=GTMDaily.

29. Peter Kelly-Detwiler, "CEO of 8minute: 'Someday Solar Energy Will Be Nearly Free,'" Forbes.com, May 1, 2019, https://www.forbes.com/sites/peterde-

twiler/2019/05/01/ceo-of-8minutenergy-someday-solar-energy-will-be-nearly-free/#2ad0070b1740.

30. In fact, a July 2020 study from the three California utilities San Diego Gas and Electric, Southern California Edison, and Pacific Gas & Electric shows that by adding four hours of storage, solar panels with trackers can be relied upon to meet peak energy demand over 99% of the time. Take away those trackers, and that number falls to 6.9%. In other words, when you really need the energy, without batteries the solar panels aren't all that good at showing up to the party. See the table in "Advice Letter," Southern California Edison Document Library, July 21, 2020, https://library.sce.com/content/dam/sce-doclib/public/regulatory/filings/pending/electric/ELECTRIC_4243-E.pdf.

31. Buttgenbach pointed to Los Angeles as an example of this new shift, noting that LA is planning to bring in a good deal of electricity from outside the basin. However, the legacy transmission system is oriented to the coast because that's where many of the thermal fossil-fueled power plants were located for access to cooling seawater. In today's world, the electricity is increasingly coming in from the east, but the transmission lines aren't built for that. "You have to take lines down and upgrade before you can bring back the system online," he commented, a process that could take 10 years. That's where strategically located solar and battery plants can make a big difference, as they take only a couple years to build.

CHAPTER 8

1. Humzah Yazdani, "Why Transmission and Distribution Are the Clean Energy Transition's Secret Weapons," World Economic Forum, July 16, 2020, https://www.weforum.org/agenda/2020/07/transmission-distribution-clean-energy-transition/.

2. "Wind Energy in Texas," American Wind Energy Association (hereafter AWEA), accessed December 6, 2020, https://www.awea.org/Awea/media/Resources/StateFactSheets/Texas.pdf.

3. "Wind Generation Growth Slowed in 2015 as Wind Speeds Declined in Key Regions," EIA, April 21, 2016, https://www.eia.gov/todayinenergy/detail.php?id=25912.

4. "Wind Energy in Texas," AWEA.

5. "USA Wind Farms File: Cactus Flats," The Wind Power, accessed November 17, 2020, https://www.thewindpower.net/windfarm_en_27851_cactus-flats.php.

6. "Wind Energy in Texas," AWEA.

7. "Wind Energy in Iowa," AWEA, accessed November 17, 2020, https://www.awea.org/Awea/media/Resources/StateFactSheets/Iowa.pdf.

8. "Wind Energy in Kansas," AWEA, accessed November 17, 2020, https://www.awea.org/Awea/media/Resources/StateFactSheets/Kansas.pdf.

9. "Wind Energy in Oklahoma," AWEA, accessed November 17, 2020, https://www.awea.org/Awea/media/Resources/StateFactSheets/Oklahoma.pdf.

10. Wind Explained: Electricity Generation from Wind," EIA, last updated March 24, 2020, https://www.eia.gov/energyexplained/wind/electricity-genera-tion-from-wind.php.

11. This is referred to in the industry as "hub height," since that's where the hub of the rotor is located.

12. Eric Lantz, Owen Roberts, Jake Nunemaker, Edgar DeMeo, Katherine Dykes, and George Scott, "Increasing Wind Turbine Tower Heights: Opportuni-ties and Challenges," National Renewable Energy Laboratory, May 2019, https://www.nrel.gov/docs/fy19osti/73629.pdf.

13. "2015 Wind Technologies Market Report," U.S. Department of Energy, August 2016, https://www.energy.gov/sites/prod/files/2016/08/f33/2015-Wind-Technologies-Market-Report-08162016.pdf.

14. Richard Kessler, "US Turbines: Bigger, Taller and Pumping Out More Power: AWEA," Recharge, updated April 17, 2020, https://www.recharge-news.com/wind/us-turbines-bigger-taller-and-pumping-out-more-power-awea/2-1-793254.

15. Ibid. Turbines are roughly 10 times larger than in the 1980s, but produce 100 times more energy.

16. "Vestas Wins First Enventus Order in USA for 336 MW Project from Ta-aleri Energia, AIP and Akuo Energy," Cision.com, June 16, 2020, https://news.cision.com/vestas-wind-systems-a-s/r/vestas-wins-first-enventus-order-in-usa-for-336-mw-project-from-taaleri-energia--aip-and-akuo-energy,c3136031.

17. Andreas Nauen, "Siemens Gamesa Renewable Energy, S.A. Inside Informa-tion," Siemens Gamesa, June 30, 2020, https://www.siemensgamesa.com/-/me-dia/siemensgamesa/downloads/en/investors-and-shareholders/inside-information-communications/2020/20200630-ip-us-eng.pdf.

18. To be somewhat pedantic, though, wind energy *is* a form of solar energy, created by combining three different dynamics: 1) the uneven heating of the earth's atmosphere; 2) irregular topography of the earth's surface; and 3) the Earth's rota-tion (the Coriolis effect—which doesn't create wind, but influences the direction).

19. "How a Wind Turbine Works," U.S. Department of Energy, June 20, 2014, https://www.energy.gov/articles/how-wind-turbine-works.

20. "How Do Wind Turbines Survive Severe Storms?" U.S. Department of Energy, accessed November 17, 2020, https://www.energy.gov/eere/articles/how-do-wind-turbines-survive-severe-storms.

21. "Size Specifications of Common Industrial Wind Turbines," AWEO.org, accessed November 17, 2020, http://www.aweo.org/windmodels.html.

22. Kevin J. Smith and Dayton Griffin, "Supersized Wind Turbine Blade Study: R&D Pathways for Supersized Wind Turbine Blades," Electricity Markets and Policy, Lawrence Berkeley Lab, March 2019, https://emp.lbl.gov/publications/supersized-wind-turbine-blade-study.

23. "Revolutionary Two-Piece Blade Design Launched, for GE Renewable Energy's Cypress Onshore Turbine Platform," LM Wind Power, accessed November 17, 2020, https://www.lmwindpower.com/en/stories-and-press/stories/news-from-lm-places/revolutionary-two-piece-blade-design-launched.

24. Nic Sharpley, "Keystone Tower Systems Installs World's First Tapered Spiral Welded Tower," Windpower Engineering & Development, May 19, 2015, https://www.windpowerengineering.com/keystone-tower-systems-installs-worlds-first-tapered-spiral-welded-tower/.

25. David Roberts, "The Tallest Wind Power Tower in the US, Assembled in One Hypnotizing Video," Vox, June 1, 2016, https://www.vox.com/2016/6/1/11820920/concrete-wind-turbine.

26. "GE Renewable Energy, COBOD and LafargeHolcim Co-Develop Record-Tall Wind Turbine Towers with 3D-Printed Concrete Bases," GE, June 17, 2020, https://www.ge.com/news/press-releases/ge-renewable-energy-cobod-and-lafargeholcim-co-develop-3D-printed-concrete-bases-wind-turbines. The partners indicated they planned to "produce ultimately a wind turbine prototype with a printed pedestal, and a production ready printer and materials range to scale up production."

27. The big question in the short term is how soon these gains will overtake the loss of the federal Production Tax Credit, which has offered a subsidy of as much as $25/MWh for the first 10 years of production.

28. "Wind Powers America," AWEA, Third Quarter 2020, https://www.awea.org/Awea/media/Resources/WPA_2020Q3_Public_Version.pdf.

29. "Electricity Explained: Electricity in the United States," EIA, last updated March 20, 2020, https://www.eia.gov/energyexplained/electricity/electricity-in-the-us.php.

30. Gerson Freitas Jr. and Chris Martin, "Cheap Wind Power Could Boost Green Hydrogen, Morgan Stanley Says," Bloomberg Quint, last updated July 27, 2020, https://www.bloombergquint.com/technology/cheap-wind-power-could-boost-green-hydrogen-morgan-stanley-says.

CHAPTER 9

1. "Bark *Wanderer* Lost," *Vineyard* [MA] *Gazette*, August 26, 1924, https://vineyardgazette.com/news/1924/08/26/bark-wanderer-lost.

2. Peter Kelly-Detwiler, "The Wind Technology Testing Center: Pushing the Envelope on Wind Technology," Forbes.com, January 9, 2013, https://www.forbes.com/sites/peterdetwiler/2013/01/09/the-wind-technology-testing-center-pushing-the-envelope-on-wind-technology/.

3. "Wind Turbines, Part 23: Full-Scale Structural Testing of Rotor Blades," International Electrotechnical Commission (IEC), April 8, 2014, https://webstore.iec.ch/publication/5436. In the world of electricity infrastructure, there is

a standard for nearly everything. The IEC 61400-23:2014 standard "defines the requirements for full-scale structural testing of wind turbine blades and for the interpretation and evaluation of achieved test results. The standard focuses on aspects of testing related to an evaluation of the integrity of the blade, for use by manufacturers and third-party investigators."

4. These enormous blades are miracles of modern engineering, from their computer-aided design to the actual methods by which they are manufactured and tested. Most blades are created by fusing two individual "shells" together. In the GE blade, about half of the material utilized is lightweight and highly durable glass fiber, while another third by weight is the resin that physically bonds the glass and other composites together to create a single-piece composite structure. That resin is injected into the blade structure through a network of machines, inlets, and over two miles of piping, within an environment closely controlled for flow rates, temperature, and humidity. The two shells are joined together, or "bonded." The 107-meter GE blades utilize over a mile of bond lines using a special glue (that dries within 30 minutes) to ensure that these joints are strong and can withstand the elements. Even the 2,000 gallons of coating/paint applied to each blade is specifically engineered and designed to protect the blade from the punishing marine environment for decades.

5. For reference, the Statue of Liberty stands at 305 feet, and the Washington Monument reaches 555 feet, while the Eiffel Tower soars to 986 feet. I had lined up a trip to the Cherbourg plant to see the joining of the blades, but COVID-19 had other plans.

6. Also referred to as the "rotor area."

7. The industry typically uses a wind speed reference of meters per second. 9.5–10.5 meters per second is optimal. As wind speeds exceed that velocity, the blades feather back to let some of the energy pass by.

8. Or the equivalent thrust for the new Aerojet Rocketdyne RS-25 engine used as the Space Shuttle's main engine.

9. According to Bill White, president and CEO of EnBW North America, and former senior director for Offshore Wind Sector Development at the Massachusetts Clean Energy Center, this was funded by the Obama-era American Recovery and Investment Act (ARRA) stimulus money and at $24 million was the largest single stimulus award. White noted that the Department of Energy had already identified the need for such a facility, and the Massachusetts congressional delegation was seeking funding support, with no solid prospects for funding. Fortunately, the project was already permitted and designed, and therefore "shovel-ready" when resources suddenly became available.

10. In May 2020, Siemens Gamesa unveiled a 108-meter blade as part of its 14 MW turbine, and almost immediately announced two sales, including one to utility Dominion Energy's 2,640 MW project off the coast of Virginia.

11. While the WTTC is currently the largest testing facility in the Americas, the United Kingdom's Catapult facility is slightly larger (it can test blades to 100 meters),

having been expanded in the last five years. See "Blades Research Hub to Support Development of Next Generation Turbine Blades," Catapult Offshore Renewable Energy, June 21, 2017, https://ore.catapult.org.uk/testing-validation/facilities/blades/. There have been recent announcements in Denmark and Germany concerning the build-out of facilities there as well, so that full blades can be tested.

12. Eric M. Hines and Mysore V. Ravindra, "Testing Tomorrow's Turbines," *Civil Engineering Magazine* 81, no. 7 (July 2011), https://ascelibrary.org/doi/10.1061/ciegag.0000359.

13. Stephanie A. McClellan, "Supply Chain Contracting Forecast for U.S. Offshore Wind Power," University of Delaware, College of Earth, Ocean, and Environment white paper, March 2019, https://cpb-us-w2.wpmucdn.com/sites.udel.edu/dist/e/10028/files/2020/01/SIOW-White-Paper-Supply-Chain-Contracting-Forecast-for-US-Offshore-Wind-Power-FINAL.pdf.

14. The International Energy Agency projects the offshore wind industry will eventually grow to be a trillion-dollar global business, theoretically capable of generating over 420,000 terawatt-hours (TWh; a terawatt is 1 trillion watts) of electricity annually; this would be 18 times more electricity than all of humanity uses today. See "Offshore Wind Outlook 2019," International Energy Agency, November 2019, https://www.iea.org/reports/offshore-wind-outlook-2019.

15. Ros Davidson, "Cape Wind: Requiem for a Dream," Wind Power Monthly, May 1, 2018, https://www.windpowermonthly.com/article/1462962/cape-wind-requiem-dream. Cape Wind—located in Nantucket Sound off Massachusetts—would have been the first American offshore wind project. After 16 years of effort and $100 million in fruitless expenditures, the initiative was scuttled, a victim of delays and over 20 lawsuits, several backed by the rich and famous living within view of the project site.

16. "Offshore Wind Ports Information Database," 4C Offshore, accessed November 18, 2020, https://www.4coffshore.com/ports/. The list of European ports capable of handling this industry numbers in the dozens—a true embarrassment of riches.

17. "Port and Infrastructure Analysis for Offshore Wind Energy Development," Massachusetts Clean Energy Center, February 2010, https://files.masscec.com/Port%20%26%20Infrastructure%20Report.pdf. It should be noted that the report assumed 3.6 MW turbines with blades "up to 197 feet," or 60 meters. At 107 meters, the Haliade-X blade is of an entirely different dimension.

18. "New Bedford Marine Commerce Terminal," Massachusetts Clean Energy Center, accessed November 18, 2020, https://www.masscec.com/facilities/new-bedford-marine-commerce-terminal.

19. Matt Miller, "New Bedford Inks Lease for Vineyard Wind Project," *American Journal of Transportation*, September 23, 2019, https://www.ajot.com/premium/ajot-new-bedford-inks-lease-for-cape-wind-offshore-wind-project. Sixty percent of that price tag was associated with environmental clean-up and remediation (the Port of New Bedford is an Environmental Protection Agency Superfund site), and

part of the effort included removal of polyvinyl chloride (PVC) as well as other contaminants when the channel was deepened.

20. "U.S. Offshore Wind Power Economic Impact Assessment," AWEA, March 2020, https://supportoffshorewind.org/wp-content/uploads/sites/6/2020/03/AWEA_Offshore-Wind-Economic-ImpactsV3.pdf. According to White, the latest totals as of July 2020 were Connecticut—2,300 MW; Maryland—1,600 MW; Massachusetts—3,200 MW; New Jersey—7,500 MW; New York—9,000 MW; Rhode Island—400 to 1,000; and Virginia—5,200 MW.

21. "Lease and Grant Information," Bureau of Ocean Energy Management (BOEM), accessed November 18, 2020, https://www.boem.gov/renewable-energy/lease-and-grant-information.

22. "Commercial Lease: Deepwater Wind New England, LLC," BOEM, October 1, 2013, https://www.boem.gov/sites/default/files/renewable-energy-program/State-Activities/RI/Executed-Lease-OCS-A-0486.pdf.

23. Commercial Lease: Equinor Wind US LLC," BOEM, April 1, 2019, https://www.boem.gov/sites/default/files/renewable-energy-program/State-Activities/MA/Lease-OCS-A-0520.pdf; "Commercial Lease: Mayflower Wind Energy LLC," BOEM, April 1, 2019, https://www.boem.gov/sites/default/files/renewable-energy-program/State-Activities/MA/Lease-OCS-A-0521.pdf; Commercial Lease; Vineyard Wind LLC," BOEM, April 1, 2-19, https://www.boem.gov/sites/default/files/renewable-energy-program/State-Activities/MA/Lease-OCS-A-0522.pdf.

24. Justin Gerdes, "Record-Breaking Massachusetts Offshore Wind Auction Reaps $405 Million in Winning Bids," GTM, December 17, 2018, https://www.greentechmedia.com/articles/read/record-breaking-massachusetts-offshore-wind-auction.

25. "Our Offshore Wind Projects in the U.S.," Ørsted, accessed November 18, 2020, https://us.orsted.com/Wind-projects.

26. Nerijus Adomaitis, "Statoil to Become Equinor, Dropping 'Oil' to Attract Young Talent," Reuters, May 15, 2018, https://www.reuters.com/article/us-statoil-agm-equinor/statoil-to-become-equinor-dropping-oil-to-attract-young-talent-idUSKCN1IG0MN.

27. "Equinor Offshore Wind Bid Wins in New York State," Equinor, July 18, 2019, https://www.equinor.com/en/news/2019-new-york-offshore-wind-bid.html.

28. Andrew Lee, "Oil Supermajor BP Backs Offshore Wind for Lift-Off," Recharge, January 22, 2020, https://www.rechargenews.com/wind/oil-supermajor-bp-backs-offshore-wind-for-lift-off/2-1-742034.

29. Adnan Durakovic, "Ørsted to Use GE Haliade-X 12 MW on US Offshore Wind Farms," OffshoreWind.biz, September 19, 2019, https://www.offshorewind.biz/2019/09/19/orsted-to-use-ge-haliade-x-12-mw-on-us-offshore-wind-farms/.

30. Sam Worley, "Big Win(d): GE's Powerful Haliade-X Offshore Wind Turbine Gets a Major U.S. Customer," General Electric, December 3, 2020, https://

www.ge.com/news/reports/big-wind-ges-powerful-haliade-x-offshore-wind-turbine-gets-a-major-us-customer.

31. In other words, there was an opportunity to reduce "balance of plant costs."'

32. This is because the energy-to-capacity ratios (capacity factors) and overall efficiencies would decline.

33. "GE Announces Haliade-X, the World's Most Powerful Offshore Wind Turbine," GE, March 1, 2018, https://www.genewsroom.com/press-releases/ge-announces-haliade-x-worlds-most-powerful-offshore-wind-turbine. In addition to decreasing the relative cost contributions of balance of the plant, the new machines boast a 63% capacity factor (defined as expected output versus 100% of what could be achieved if running every hour of the year at full capacity), a record for the wind industry.

34. It's a highly competitive game, indeed. Less than a month after our conversation, rival Siemens Gamesa announced a contract with Dominion Energy for its $7.8 billion, 2,640 MW project off Virginia's shores (Karl-Erik Stromsta, "Siemens Gamesa Wins 2.6GW Offshore Turbine Deal for Dominion's Virginia Project," GTM, January 7, 2020, https://www.greentechmedia.com/articles/read/largest-us-offshore-wind-taps-siemens-gamesa-for-turbine-order). That project is likely to utilize Siemens Gamesa's digitally enhanced 11 MW turbine. However, this machine is meant to bridge the demand until it brings on what its CEO calls a "true step change" in its 1X platform expected to be available by 2024–2025. See Darius Snieckus, "Siemens Gamesa Unveils Digitally Souped-Up 11 MW Offshore Turbine," Recharge, November 26, 2019, https://www.rechargenews.com/wind/siemens-gamesa-unveils-digitally-souped-up-11mw-offshore-turbine/2-1-711795. In other words, the arms race continues.

35. In fact, Ørsted recently admitted that it had dramatically underestimated the impacts of turbines on overall output, both within existing arrays as well as between various installations. It noted that it has also underestimated the "blocking effect" that occurs when the wind actually slows down as it reaches the turbine. This reduces each individual turbine's output while having an even larger cumulative effect on the entire wind farm. The "wake effect," both within and between adjacent wind farms, occurs as a result of the wind slowing down after passing through the turbine blades, having been affected similar to the way wind spills off a boat's sail when heading into the wind. Though this effect has been modeled for years, Ørsted has stated, "Our results point to a higher negative effect on production that earlier models have predicted," and commented, "We believe that underestimation of blockage and wake effects is likely to be an industry-wide issue." The company reduced its financial targets for some offshore wind farms as a consequence. See "Ørsted Presents Update on Its Long-Term Financial Targets," Ørsted, October 29, 2019, https://orsted.com/en/Company-Announcement-List/2019/10/1937002.

36. "Report: U.S. Offshore Wind a $70 Billion Opportunity for Supply Chain," Maritime Executive, March 26, 2019, https://www.maritime-executive.com/article/report-u-s-offshore-wind-a-70-billion-opportunity-for-supply-chain.

37. Nicholas Pugliese, "N.J. Coal Plant Retiring, Imperiling Pipeline Plan While Boosting Offshore Wind," WHYY, May 1, 2019, https://whyy.org/articles/nj-coal-plant-retiring-imperiling-pipeline-plan-while-boosting-offshore-wind/.

38. "Massachusetts Agency Awards Vineyard Wind Key Permit for Construction of Offshore Wind Farm Interconnection to Regional Grid," Vineyard Wind, May 9, 2019, https://www.vineyardwind.com/press-releases/2019/5/9/massachusetts-agency-awards-vineyard-wind-key-permit-for-construction-of-offshore-wind-farm-interconnection-to-regional-grid.

39. "The Neptune Project," Anbaric, accessed November 18, 2020, https://anbaric.com/neptune/.

40. Herman K. Trabish, "Texas CREZ Lines Delivering Grid Benefits at $7B Price Tag," Utility Dive, June 25, 2014, https://www.utilitydive.com/news/texas-crez-lines-delivering-grid-benefits-at-7b-price-tag/278834/; "Texas Is the National Leader in the U.S. Wind Energy Industry," AWEA, accessed November 18, 2020, https://www.awea.org/Awea/media/Resources/StateFactSheets/Texas.pdf. With an installed capacity of 29,400 MWs in the first quarter of 2020, Texas as a country would be fifth in wind, trailing only China, the United States, Germany, and India. See "World's Top 10 Countries in Wind Energy Capacity," Energyworld.com, March 18, 2019, https://energy.economictimes.indiatimes.com/news/renewable/worlds-top-10-countries-in-wind-energy-capacity/68465090.

41. "Southern New England OceanGrid," Anbaric, accessed November 19, 2020, https://anbaric.com/southernnewenglandoceangrid/.

CHAPTER 10

1. "California Household Battery Sales to Quadruple in 2020," Bloomberg NEF, February 10, 2020, https://about.bnef.com/blog/california-household-battery-sales-to-quadruple-in-2020/.

2. Ivan Penn and Peter Eavis, "PG&E Pleads Guilty to 84 Counts of Manslaughter in Camp Fire Case," *New York Times*, June 16, 2020, https://www.nytimes.com/2020/06/16/business/energy-environment/pge-camp-fire-california-wildfires.html.

3. Janie Har, "PG&E Trying a New Tactic in Power Cuts to Prevent Wildfires," Powergrid International, September 9, 2020, https://www.power-grid.com/2020/09/09/pge-trying-a-new-tactic-in-power-cuts-to-prevent-wildfires/?utm_medium=email&utm_campaign=2020-09-10&utm_source=powergrid_weekly_newsletter.

4. Peter Kelly-Detwiler, "How NEC Energy Solutions Is Enabling the Batter Revolution," Forbes.com, January 16, 2020, https://www.forbes.com/sites/peterdetwiler/2020/01/16/a-battery-of-opportunities-nec-energy-solutions-looks-to-the-future-of-energy-storage/#6f3501661ade.

5. In mid-June, the company shuttered its doors, citing the pandemic and resulting economic disruption for its failure to find a buyer. With nearly 1,000 MW of projects either in the field or in development, it was one of the larger players, but had yet to turn a profit. David R. Baker and Brian Eckhouse, "A Global Battery Firm Is Going Out of Business, Citing Covid-19," Bloomberg, last updated June 12, 2020, https://www.bloomberg.com/news/articles/2020-06-12/a-global-battery-firm-is-going-out-of-business-citing-covid?sref=xfyiavTX.

6. The lithium-ion battery has been around and in commercial use since 1991, when Sony first used it in its video camcorder. In the ensuing years, variants of that chemistry were incorporated into other electronic devices, eventually dominating the consumer electronics segment. Lithium-ion battery technology then found its way to EVs, largely because it packed a solid punch in terms of energy density—the ability to delivery power (how much can be released instantaneously) and energy (how much can be released over a given duration) relative to the weight of the battery. However, over time, a number of different lithium-ion technologies emerged, and each had its own characteristics that made the battery better suited for some applications than for others.

It may help to briefly understand battery basics. Batteries consist of an anode, a cathode, and an electrolyte separating them. When the battery is charged with an electric current, positive lithium ions (lithium atoms lacking an electron) travel from the cathode to the anode through the electrolyte. In laymen's terms, the lithium ions would rather be nestled in among the cathode particles than the anode carbon particles, but they can't move through the electrolyte as an atom, only as an ion. So when an external circuit such as a motor or light bulb discharges the battery by accepting electrons from the anode through their circuits, the lithium ions happily move through the electrolyte back to the cathode and meet up with their electrons there. The driving force that causes the ions to "want" to get back to the cathode also creates the voltage of the cell. The difference between the ionic attractiveness of different materials causes cells with different chemistries to exhibit different characteristic voltages.

If that's an inexplicable mystery to you, no matter. What's really important is that the composition of the cathode particles is the secret sauce that imparts the essential characteristics to different lithium battery chemistries. There are two main chemistries currently being deployed on the grid, growing out of applications in the electrified transportation industry, or "e-mobility." These differ slightly with respect to a few key properties: cost, energy density, shelf life, cycle life, and stability.

Shelf life is defined as how useful the battery is after being stored for a given period of time. Cycle life refers to how often one can charge and discharge a battery before it can no longer deliver its desired power and/or energy. And stability relates to the ability of the battery to withstand abuse before it ignites—a critical issue when one considers that these devices are being placed in our cars, our office buildings, and our homes.

7. Two of the most common lithium-ion technologies used in the grid are lithium-iron-phosphate (LFP) and nickel-manganese-cobalt (NMC). LFP is less energy-dense than other lithium technologies. However, it is also less expensive to manufacture, has a relatively long cycle life (over 5,000 cycles), and good stability (its thermal runaway onset temperature is 270 degrees C (518 degrees F). It has been the predominant technology for more than 90% of China's 420,000 buses that operate in congested areas, and don't travel at either high speed or for long distances. The lithium iron phosphate chemistry is more frequently deployed in stationary storage applications where density is less of an issue (one is not carrying the battery around on wheels, so weight doesn't matter), and stability is of critical importance. NMC also has relatively high energy densities and is used in many EVs, such as the Nissan Leaf, Chevrolet Bolt, BMW i3, and VW's many new models. EV versions of this chemistry offer approximately 1,000–2,000 cycles until they reach 80% of their original capacity, and have a thermal runaway onset temperature around 210 degrees C (410 degrees F). NMC batteries designed for grid applications regularly achieve over 5,000 cycles (sometimes only partial cycles) until they are no longer usable in that application. This chemistry has been changing a good deal in recent years as manufacturers learn how to improve cost, energy density, cycle life, and charge rates as well as use less controversial materials such as cobalt.

Cobalt helps to stabilize the chemistry, but it is also known for its unsavory origins, with roughly 60% emanating from the Democratic Republic of the Congo, where a measurable amount of supply has been associated with child labor. It's also difficult to quickly ramp up the supply of cobalt in response to rising demand. Almost all cobalt is mined in association with copper or nickel coming from the same ore body, and is not generally economical to mine on its own merits. As a consequence, the prices have been very fickle, rising and falling by as much as 100% or more within short timeframes. In response, a number of companies have been striving to reduce the quantity of cobalt in the battery mix, while others are working to create ethically sourced cobalt.

For its part, lithium is sourced largely from the salt flats in the Altiplano lying between Chile, Argentina, and Bolivia. Its mining and processing has created some environmental concerns, largely a consequence of the required water use in a very arid region of the world. Lithium is also sourced from hard rock spodumene mines in Australia.

8. "Tesla Gigafactory," Tesla, accessed November 19, 2020, https://www.tesla.com/gigafactory.

9. John Parnell, "A Battery Giant Is Born: Total and Groupe PSA Launch New European Manufacturer," GTM, September 4, 2020, https://www.greentechmedia.com/articles/read/a-battery-giant-is-born-total-and-groupe-psa-launch-new-company.

10. CATL Cars: Germany's Big Chinese-Made Battery Plant to Dwarf Tesla's Gigafactory," Handelsblatt, February 6, 2019, https://www.handelsblatt.com/english/companies/catl-cars-germanys-big-chinese-made-battery-plant-to-dwarf-teslas-gigafactory/23955856.html?ticket=ST-8280192-vZrJkhQ0OZPj5itX5duf-ap4.

11. "This Chinese Battery Maker Hopes to Power Up the Global Electric Car Market," Fortune.com, December 25, 2016, https://fortune.com/2016/12/25/catl-chinese-electric-car-battery-maker/.

12. "New Lab to Develop Next-Generation Batteries," Xinhuanet, June 25, 2020, http://www.xinhuanet.com/english/2020-06/25/c_139166464.htm.

13. Ibid.

14. Norihiko Shirozu and Paul Lienert, "Exclusive: Tesla's Secret Batteries Aim to Rework the Math for Electric Cars and the Grid," Reuters, May 14, 2020, https://www.reuters.com/article/us-autos-tesla-batteries-exclusive/exclusive-teslas-secret-batteries-aim-to-rework-the-math-for-electric-cars-and-the-grid-idUSKBN22Q1WC.

15. "A Million-Mile Battery from China Could Power Your Electric Car," Bloomberg News, June 7, 2020, https://www.bloomberg.com/news/articles/2020-06-07/a-million-mile-battery-from-china-could-power-your-electric-car?sref=xfyiavTX.

16. There are other lithium technologies—such as lithium titanate—that deliver many more cycles, but they are cost-prohibitive.

17. In the United States alone, over $7 billion in storage projects are expected to be installed from the period 2021–2025, though that figure may be somewhat impacted by COVID-19-related slowdowns. See "Webinar: Grid-Connected Energy Storage Supply Chains in the Era of Coronavirus," Energy Storage Association, April 8, 2020, https://energystorage.org/wp/wp-content/uploads/2020/04/Grid-Connected-ES-Supply-Chain_COVID-19-Slides-Final.pdf.

18. On a July conference call I attended, an executive from one of the larger renewables and storage developers in the country stated, "Any storage project we build today will change technology six or seven years from now."

19. Projects are generally modular and portable. In fact, many units are actually housed in shipping containers (although newer and smaller units are increasingly being daisy-chained together in purpose-built shells). For its part, NEC had shipping containers in three different sizes—20, 40, and 53 feet—that could be replicated on-site with as many units as needed, with the largest container supporting 5.5 MWh of energy and about 9 MW of power (the amount of electricity required at any one moment).

20. "Why Choose 8minute?" 8minute, accessed November 20, 2020, https://www.8minute.com.

21. "Vistra Considering 1,500 MW/6,000 MWh Battery Storage Complex in California," Institute for Energy Economics and Financial Analysis (IEEFA), August 12, 2020, https://ieefa.org/vistra-considering-1500mw-6000mwh-battery-storage-complex-in-california/. For context, total U.S. installations of storage capacity were 311 MW in 2018 and 523 MW in 2019.

22. "Glenwood Management Earns Demand Response Revenue, Reduces Energy Costs with Intelligent Energy Storage," Enel X, accessed November 20, 2020, https://www.enelx.com/n-a/en/resources/case-study/cs-glenwood-management.

23. There were 546 microgrids installed in the United States in 2019. Tim Sylvia, "The U.S. Installed More Microgrids in 2019 Than Ever Before," *PV Magazine*, July 25, 2020, https://pv-magazine-usa.com/2020/07/25/the-u-s-installed-more-microgrids-in-2019-than-ever-before/.

24. Conor Ryan, "Ameresco Completes 8MWh Battery Storage System at US Marine Corps Recruit Depot," Energy Storage News, June 20, 2019, https://www.energy-storage.news/news/ameresco-completes-8mw-battery-storage-system-at-us-marine-corps-recruit-de.

25. Darrell Proctor, "Power Project Provides Shelter from the Storm," Power, June 1, 2020, https://www.powermag.com/power-project-provides-shelter-from-the-storm/.

26. "McMicken Investigation," APS, July 27, 2020, https://www.aps.com/en/About/Our-Company/Newsroom/Articles/Equipment-failure-at-McMicken-Battery-Facility. Ironically, even with all that focus on safety, one of NEC's installations in Liverpool, England, suffered a fire in September 2020. As of this writing, the cause was still unknown. See Andy Colthorpe, "Fire at 20MW UK Battery Storage Plant in Liverpool," Energy Storage News, November 25, 2020, https://www.energy-storage.news/news/fire-at-20mw-uk-battery-storage-plant-in-liverpool.

27. The explosion was caused by a buildup of explosive gases from the cells undergoing thermal runaway. To mitigate the danger due to the accumulation of explosive gases, the explosive gases must be removed from the container before the atmosphere in the container reaches its lower flammability level (LFL).

28. "McMicken Battery Energy Storage System Event Technical Analysis and Recommendations," Arizona Public Service, July 18, 2020, https://www.aps.com/-/media/APS/APSCOM-PDFs/About/Our-Company/Newsroom/McMickenFinalTechnicalReport.ashx?la=en&hash=50335FB5098D9858BFD276C40FA54FCE. It's for that safety reason that the New York City Fire Department (FDNY) still will not permit the deployment of lithium-ion batteries within buildings, although most other cities do not appear to have such concerns. There are a few battery systems in the city, but they use a more stable alternative chemistry, such as lead acid batteries.

29. Julian Spector, "How the Energy Storage Industry Responded to the Arizona Battery Fire," GTM, August 18, 2020, https://www.greentechmedia.com/articles/read/arizona-battery-fire-already-prompted-safety-improvements-in-grid-storage.

30. Hauke Engel, Patrick Hertzke, and Giulia Siccardo, "Breathing New Life into Used Electric Vehicle Batteries," McKinsey & Company, July 9, 2019, https://www.mckinsey.com/business-functions/sustainability/our-insights/sustainability-blog/breathing-new-life-into-used-electric-vehicle-batteries#.

31. Mark Kane, "148 Nissan LEAF Batteries Power This Stadium," Inside EVs, June 30, 2018, https://insideevs.com/news/338994/148-nissan-leaf-batteries-power-this-stadium/.

32. "UL Issues World's First Certification for Repurposed EV Batteries to 4R Energy," UL, August 19, 2019, https://www.ul.com/news/ul-issues-world%27s-first-certification-repurposed-ev-batteries-4r-energy.

33. "Levelized Cost of Energy and Levelized Cost of Storage 2019," Lazard, November 7, 2019, https://www.lazard.com/perspective/lcoe2019. Financial firm Lazard LLC publishes a yearly levelized cost of energy report that assess the levelized costs per megawatt-hour for various energy resources. Those costs include the upfront capital, interest, operations and maintenance, and decommissioning at the end of the project lifetime.

34. "Advice Letter Suspension Notice: Energy Division," Southern California Edison, July 21, 2020, https://library.sce.com/content/dam/sce-doclib/public/regulatory/filings/pending/electric/ELECTRIC_4243-E.pdf.

35. "Pumped-Storage Hydropower," U.S. Department of Energy, Office of Energy Efficiency and Renewable Energy, accessed November 21, 2020, https://www.energy.gov/eere/water/pumped-storage-hydropower.

36. "2018 Pumped Storage Report," National Hydropower Association, April 2018, https://www.hydro.org/wp-content/uploads/2018/04/2018-NHA-Pumped-Storage-Report.pdf.

37. "Cryogenic Energy Storage," Highview Power, accessed November 21, 2020, https://www.highviewpower.com/technology/. The first U.S. project was announced by Highview Power in late 2019, and will involve 50 MW and eight hours of storage located in northern Vermont. The installation will improve the economics of a local wind farm that cannot deliver energy into a highly "congested" overloaded grid during parts of the day. At present, that wind energy is simply curtailed—in other words, wasted. That energy, whose cost during those hours is essentially free, will be stored and delivered into the grid during periods when the transmission system is less taxed and/or when the energy is needed.

38. Molly Seltzer, "Why Salt Is This Power Plant's Most Valuable Asset," *Smithsonian Magazine*, August 4, 2017, https://www.smithsonianmag.com/innovation/salt-power-plant-most-valuable-180964307/.

39. "Advanced Compressed Air Energy Storage: A Unique Grid Storage Solution," Hydrostor, accessed November 21, 2020, https://www.hydrostor.ca/technology/.

40. "About Us," Energy Vault, accessed November 21, 2020, https://energyvault.com.

41. "Form Energy Announces Pilot with Great River Energy," *Renewable Energy Magazine*, May 11, 2020, https://www.renewableenergymagazine.com/storage/form-energy-announces-pilot-with-great-river-20200511.

CHAPTER 11

1. In 2019, gas turbines were responsible for 38.4% of total electricity generation, followed by coal at 23.4%, nuclear at 19.6%, and renewables at 17.6%. "What

Is U.S. Electricity Generation by Energy Source?" EIA, last updated November 2, 2020, https://www.eia.gov/tools/faqs/faq.php?id=427&t=6.

2. "U.S. Electric Generating Capacity Increase in 2016 Was Largest Net Change Since 2011," IEA, February 27, 2017, https://www.eia.gov/todayinenergy/detail.php?id=30112.

3. America's Electricity Generating Capacity," American Public Power Association, accessed November 22, 2020, https://www.publicpower.org/resource/americas-electricity-generating-capacity.

4. "Natural Gas at a Glance," Center for Climate and Energy Solutions, accessed November 22, 2020, https://www.c2es.org/content/natural-gas/.

5. Bobby Weaver, "Turning to the Right," Permian Basin Petroleum Association Magazine, January 1, 2017, https://pboilandgasmagazine.com/turning-to-the-right/.

6. Many of these are known but not disclosed, and are considered hazardous to human health.

7. Stephen Yang, "Hydraulic Fracturing—Potential for Contamination of Drinking Water Sources," *State of the Planet* (blog), Columbia University, May 3, 2010, https://blogs.ei.columbia.edu/2010/05/03/hydraulic-fracturing-potential-for-contamination-of-drinking-water-sources/.

8. "How Much Water Does the Typical Hydraulically Fractured Well Require?" U.S. Geological Survey, accessed December 1, 2020, https://www.usgs.gov/faqs/how-much-water-does-typical-hydraulically-fractured-well-require?qt-news_science_products=0#qt-news_science_products.

9. "Frac Water Demand Is Sky-Rocketing," Rystad Energy, January 22, 2019, https://www.rystadenergy.com/newsevents/news/press-releases/Frac-water-demand-is-sky-rocketing/.

10. A. Pena Castro, Sara L. Dougherty, R. M Harrington, and Elizabeth Cochran, "Delayed Dynamic Triggering of Disposal-Induced Earthquakes Observed by a Dense Array in Northern Oklahoma," *Journal of Geophysical Research B: Solid Earth*, 124 (2019), U.S. Geological Survey, https://pubs.er.usgs.gov/publication/70203264.

11. "More Than 100 Coal-Fired Plants Have Been Replaced or Converted to Natural Gas Since 2011," EIA, August 5, 2020, https://www.eia.gov/todayinenergy/detail.php?id=44636.

12. Capacity factors are simply utilization rates, defined as the percentage of time a plant operates compared with the amount of time it could operate if it were to run 24/7, 365 days a year. It's like an airplane with 250 seats, 200 of which are occupied by passengers; it would have an 80% capacity factor for that flight. A higher capacity factor implies higher revenues and generally a more profitable operation.

13. Michael Hawthorne, "Without Another Bailout, Exelon Plans to Close Two Illinois Nuclear Plants Next Year," *Chicago Tribune*, August 28, 2020, https://www.chicagotribune.com/news/environment/ct-exelon-nuclear-plants-shut-down-20200828-qmk6z3d5mrgipeahb56vugq3za-story.html.

14. "Five States Have Implemented Programs to Assist Nuclear Power Plants," EIA, October 7, 2019, https://www.eia.gov/todayinenergy/detail.php?id=41534.

15. "Renewable Energy Providing a Record 40 Percent of All Oklahoma Power," Oklahoma Power Alliance, March 4, 2020, https://okpoweralliance.com/uncategorized/renewable-energy-providing-a-record-40-percent-of-all-oklahoma-power/.

16. "State Solar Spotlight: California," Solar Energy Industries Association, December 10, 2019, https://www.seia.org/sites/default/files/2019-12/California.pdf This figure is through the third quarter of 2019.

17. "Hawaii Solar," Solar Energy Industries Association, accessed November 22, 2020, https://www.seia.org/state-solar-policy/hawaii-solar. This figure is through the third quarter of 2019.

18. "US Wind Industry Delivers Strong First Quarter," AWEA, April 20, 2020, https://www.awea.org/resources/news/2020/us-wind-industry-delivers-strong-first-quarter. The AWEA counted 107,443 turbines dotting the American landscape as of April 2020.

19. "What Are Atmospheric Rivers?" National Oceanic and Atmospheric Administration, last updated December 2015, https://www.noaa.gov/stories/what-are-atmospheric-rivers.

20. The duck chart also illustrates a rate of change that caught planners entirely by surprise: the original estimates laid out in 2012 when this duck was originally hatched were that net generation requirements would fall to about 12,000 MW by 2020. The reality was that the duck grew a lot fatter, a whole lot faster than anybody expected, as investors and homeowners alike poured into the solar space. Two years ahead of the 2020 date, actual net load on a day in February fell to under 7,000 MW.

21. Nathaniel Bullard, "New California Study Is Good News for Certain Solar Companies," Bloomberg News, July 16, 2020, https://www.bloomberg.com/news/articles/2020-07-16/new-california-study-is-good-news-for-certain-solar-companies?sref=xfyiavTX.

22. "About Western Energy Imbalance Market," Western Energy Imbalance Market, accessed November 22, 2020, https://www.westerneim.com/Pages/About/default.aspx.

23. This dynamic probably won't slow down anytime soon, although it may change a little with the addition of energy storage, which will allow solar resources to shift the timing of their energy deliveries into the grid. As of January 2020, all new homes in California are required to be equipped with rooftop solar.

24. "Advice Letter Suspension Notice: Energy Division, Southern California Edison, July 21, 2020, https://library.sce.com/content/dam/sce-doclib/public/regulatory/filings/pending/electric/ELECTRIC_4243-E.pdf.

25. Supercomputers have been used to model heat flows inside the turbine, and push machines to new frontiers. For example, GE Research was recently awarded access to utilize Oak Ridge National Laboratory's 200-petaflop supercomputer (a

petaflop is 1 quadrillion calculations per second). This machine, called Summit, will allow GE researchers to accelerate the process of innovation by better understanding the operation of complex industrial systems in ways that were previously impossible. See "GE Research Leverages DOE's Summit Supercomputer to Boost Jet Engine and Power Generation Equipment Efficiency," HPC Wire, February 19, 2020, https://www.hpcwire.com/off-the-wire/ge-research-leverages-does-summit-supercomputer-to-boost-jet-engine-and-power-generation-equipment-efficiency.

26. Aiden Burgess, "GE Awarded World Record for Most Efficient Combined-Cycle Power Plant," The Manufacturer, June 27, 2016, https://www.themanu-facturer.com/articles/ge-world-record-for-most-efficient-combined-cycle-power-plant/. For example, GE's HA turbine in its 630 MW Bouchain plant—which set the world record for efficiency, converting 62.22% of the fuel energy contained in natural gas into power in 2016—burns at the high end of that range.

27. Ibid. GE indicates it could fill the Goodyear Blimp in 10 seconds.

28. In North America, with our grid frequency of 60 Hz, turbines spin at 3,600 rpm. In much of the rest of the world governed by 50 Hz, turbines spin at 3,000 rpm.

29. The Southeastern United States was no stranger to that scenario in the winter of 2019–2020, and California frequently experiences atmospheric rivers in which huge fronts coming off the Pacific result in rain for days on end. Cloud cover can reduce solar output by 50% or more. With solar panels on trackers, rather than fixed-tilt arrays that usually sit at an optimum angle depending on the plant's latitude, the panels can be set to horizontal mode to capture all of the dispersed irradiance.

30. Nuclear plants exist in the generation mix in many parts of the country as well, but they are highly inflexible and very difficult to ramp up and down.

31. "9HA.01/.02 Gas Turbine (50 Hz)," GE Power, accessed November 22, 2020, https://www.ge.com/power/gas/gas-turbines/9ha.

32. "Breaking the Power Plant Efficiency Record: GE & EDF Unveil a Game-Changer at Bouchain," GE Power, April 2016, https://www.ge.com/power/about/insights/articles/2016/04/power-plant-efficiency-record.

33. Tomas Kellner, "Here's Why the Latest Guinness World Record Will Keep France Lit Up Long After Soccer Fans Leave," GE, June 17, 2016, https://www.ge.com/reports/bouchain/.

34. Amy Kover, "Oops! They Did It Again: GE Turbine Delivers Second World Record," GE, April 24, 2018, https://www.ge.com/reports/whoops-ge-turbine-delivers-second-world-record/.

35. Darrell Proctor, "Efficiency Improvements Mark Advances in Gas Turbines," Power, January 3, 2018, https://www.powermag.com/efficiency-improve-ments-mark-advances-in-gas-turbines/.

36. Peter Kelly-Detwiler, "Timing Your Power Plant Outage Schedule for Enhanced Efficiency and Profitability," Transform, March 6, 2017, https://www.ge.com/power/transform/article.transform.articles.2017.mar.timing-your-power-plant-outage.

CHAPTER 12

1. Emissions are defined as falling into three separate categories, or "scopes." Scope 1 emissions are those directly emitted by a company's activities and facilities, such as vehicle fuel and boilers combusting fuel on-site. Scope 2 emissions come from emissions associated with electricity consumption, and Scope 3 emissions are associated with emissions in related supply chains. See "What Is the Difference Between Scope 1, 2 and 3 Emissions?" Compare Your Footprint, November 2, 2018, https://compareyourfootprint.com/difference-scope-1-2-3-emissions/.

2. "RE100 Members," RE100 Climate Group, accessed December 8, 2020, https://www.there100.org/companies. Many of these are large international companies, ranging from GM to IKEA to Mars, Inc.—the first U.S. member.

3. "Corporate Clean Energy Buying Leapt 44% in 2019, Sets New Record," Bloomberg NEF, January 28, 2020, https://about.bnef.com/blog/corporate-clean-energy-buying-leapt-44-in-2019-sets-new-record/.

4. "BRC Deal Tracker," Business Renewables Center, accessed November 23, 2020, https://businessrenewables.org/corporate-transactions/.

5. In the fall of 2019, for example, northern California utility Pacific Gas & Electric implemented preemptive "Public Power Safety Shutdowns" to mitigate the risk of fire. These affected hundreds of thousands of individuals. See "PG&E Slammed for Cutting Power to Millions of Californians," NBC News, October 10, 2019, https://www.nbcnews.com/news/us-news/pg-e-slammed-cutting-power-millions-californians-n1064481. The summer of 2020 saw similar activities.

6. "Novartis Announces US Renewables Agreement," Novartis, accessed November 23, 2020, https://www.novartis.us/news/media-releases/novartis-announces-us-renewables-agreement-reduce-greenhouse-gas-emissions.

7. The Zaragosa project covered approximately 2 million square feet of roof space on the plant with about 85,000 solar panels; see "GM Adding World's Largest Rooftop Solar Power System to Plant," Air Conditioning/Heating/Refrigeration News, July 25, 2008, https://www.achrnews.com/articles/107739-july-25-2008-gm-adding-world-s-largest-rooftop-solar-power-system-to-plant.

8. In a rare and astounding exception, datacenter company Switch announced in July 2020 "the world's largest behind-the-meter solar project in the world," at its location in Storey County, Nevada, at 127 MW and combined with a 240 MWh battery storage system.

9. Alyssa Danigelis, "GM Takes the Long View on Emissions Reductions: Q&A with Dane Parker," Environment+Energy Leader, March 31, 2020, https://www.environmentalleader.com/2020/03/gm-long-view-emissions-qa/.

10. "General Motors and DTE Energy Are Making Michigan a Clean Energy Powerhouse," GM, April 20, 2020, https://media.gm.com/media/us/en/gm/news.detail.html/content/Pages/news/us/en/2020/apr/0420-migp.html.

11. "State Renewable Portfolio Standards and Goals," National Conference of State Legislators, April 17, 2020, https://www.ncsl.org/research/energy/renewable-portfolio-standards.aspx.

12. Personally, I did just that. I sold the "solar renewable energy credits" or SRECs for the output of five panels that I own to a third party that took credit for them. I received approximately $.50 per SREC, well over twice what I made from the value of the actual energy generated and credited to my electricity bill. Without SRECs, I would not have entered the deal.

13. The EPA states that "Green power procured for a 12-month reporting period must be either generated during that reporting period, generated during the six (6) months immediately preceding the reporting period, or the three (3) months immediately following the reporting period. This equates to a 21-month eligibility period for which a Partner's renewable energy certificate-based green power can be generated." See "Green Power Partnership Eligible Generation Dates," U.S. Environmental Protection Agency, accessed December 6, 2020, https://www.epa.gov/greenpower/green-power-partnership-eligible-generation-dates.

14. Jordan Larimore, "Billion-Dollar Wind Farm Focus of Public Hearing," *Joplin Globe,* January 19, 2019, https://www.joplinglobe.com/news/local_news/billion-dollar-wind-farm-focus-of-public-hearing/article_27e70a5c-2133-51dc-9da6-48a163e0c948.html.

15. As LevelTen Energy comments, "in the best-case scenario, a PPA delivers a massive amount of RECs for a corporation and actually earns the corporation money. In the worst-case scenario, the corporation ends up paying much more than they expected each month, or the project doesn't get built at all, which means no RECs would be delivered." Jessica Johnson, "4 Ways to Get Renewable Energy Certificates: Pros & Cons of Each," LevelTen Energy, May 21, 2020, https://leveltenenergy.com/blog/clean-energy-experts/ways-to-get-renewable-energy-certificates/.

16. Google, which pioneered the first fixed-floating swap VPPA in 2010, observed in a 2016 white paper on the topic: "For all the benefits of fixed-floating swaps, however, the model also creates unnecessary layers of complexity and dilutes the financial benefits that we receive as an end user. Because of restrictive retail market structures, we are essentially buying power twice and selling it once—buying once at the competitive wholesale level and again at the regulated retail level, while we also sell at the competitive wholesale level. Since these two prices aren't always correlated, we don't reduce our exposure to market price volatility quite as much. Further, these structures also require significant resources and expertise to execute, as well as a long-term commitment from the buyer." See "Achieving Our 100% Renewable Energy Purchasing Goal and Going Beyond," Google, December 2016, https://static.googleusercontent.com/media/www.google.com/en//green/pdf/achieving-100-renewable-energy-purchasing-goal.pdf.

17. In early 2018, McDonalds was the first restaurant company to establish science-based reduction targets to cut greenhouse gas emissions from its offices and restaurants by 36% by 2030 (relative to 2015 levels). It also committed to emissions reductions of 31% across its entire supply chain by 2030. Robert Gibbs, "Using Our Scale for Good," McDonalds, accessed December 6, 2020, https://corporate.

mcdonalds.com/corpmcd/en-us/our-stories/article/ourstories.carbon_footprint. html.

18. "6 Risks You Need to Understand Before Entering into a Power Purchase Agreement," LevelTen Energy, January 24, 2020, https://leveltenenergy.com/ blog/ppa-risk-management/power-purchase-agreement-risks/.

19. Peter Kelly-Detwiler, "Solar Technology Will Just Keep Getting Better: Here's Why," Forbes.com, September 26, 2019, https://www.forbes.com/sites/ peterdetwiler/2019/09/26/solar-technology-will-just-keep-getting-better-heres-why/#617f30287c6b.

20. "Lazard's Levelized Cost of Energy Analysis: Version 13.0," Lazard, 2019, https://www.lazard.com/media/451086/lazards-levelized-cost-of-energy-version-130-vf.pdf.

21. Sarah Golden, "Wind Deals Are Becoming Even More Popular with Corporate Renewables Buyers," GreenBiz, June 19, 2020, https://www.greenbiz. com/article/wind-deals-are-becoming-even-more-popular-corporate-renewables-buyers.

22. "ERCOT Fact Sheet," Electric Reliability Council of Texas, January 2020, http://www.ercot.com/content/wcm/lists/197391/ERCOT_Fact_ Sheet_1.15.20.pdf.

23. "Wind Integration Reports," ERCOT, accessed November 23, 2020, http://www.ercot.com/gridinfo/generation/windintegration.

24. That's a peculiar aspect of renewable assets in general. Since they are non-dispatchable and rely on the same available free fuel—whether in solar or wind form—their value (yes, the actual term is economic utility) begins to decline as more and more new units are added. It's a poor analogy, but an analogy nonetheless: the first beer is often thirst-quenching and remarkable. The fourth is OK. And anything after six results in negative utility.

25. "Achieving Our 100% Renewable Energy Purchasing Goal and Going Beyond," Google, December 2016, https://static.googleusercontent.com/media/ www.google.com/en//green/pdf/achieving-100-renewable-energy-purchasing-goal.pdf.

26. Ana Radovanovic, "Our Data Centers Now Work Harder When the Sun Shines and Wind Blows," Google, April 22, 2020, https://www.blog.google/ inside-google/infrastructure/data-centers-work-harder-sun-shines-wind-blows/.

27. In some ways, the actual concept is not new. The Demand Response team I headed at Constellation was working with a large data company in 2008 and 2009 to create a concept we called "follow the moon." It was essentially a strategy to shift elastic data center loads and migrate them across the globe to wherever the price of power was cheapest. At the time, we abandoned the concept as being too difficult to implement. Although it made sense economically, since one could continuously arbitrage power across the planet, it was too difficult to achieve with the available resources. There were, quite simply, easier ways to make money.

28. Lucas Joppa, "Progress on Our Goal to Be Carbon Negative by 2030," Microsoft, July 21, 2020, https://blogs.microsoft.com/on-the-issues/2020/07/21/carbon-negative-transform-to-net-zero/.

29. Ibid.

CHAPTER 13

1. One can buy the Ioniq as an internal combustion vehicle, a battery hybrid, or all electric.

2. After I leased the car, I did a lot of research to find out more about the deal and how prevalent it was. I had friends in other states who were unable to find it. As it turned out, it was good enough to elicit the interest of journalists; see Bengt Halvorson, "At $79 a Month, Is This the Best Electric Car Lease Deal of 2019?" Green Car Reports, November 13, 2019, https://www.greencarreports.com/news/1125981_at-79-a-month-is-this-the-best-electric-car-lease-deal-of-2019. But the comments in the articles pretty much said the same thing—there were no such leases to be had. It's likely only a few dozen such leases were made available.

3. "2020 Nissan LEAF Range, Charging & Battery," Nissan, accessed November 24, 2020, https://www.nissanusa.com/vehicles/electric-cars/leaf/features/range-charging-battery.html.

4. "Bolt EV," Chevrolet, accessed November 24, 2020, https://www.chevrolet.com/electric/bolt-ev.

5. "Tesla: Select Your Car," Tesla, accessed November 24, 2020, https://www.tesla.com/model3/design#battery. The June 2020 base price, before incentives, is $37,990, and as low as $31,190 after "potential savings." It costs an additional $9,000 for the extra 72 miles of range, but that also comes with other enhancements such as dual all-wheel drive.

6. "Model S Long Range Plus: Building the First 400-Mile Electric Vehicle," Tesla, June 15, 2020, https://www.tesla.com/blog/model-s-long-range-plus-building-first-400-mile-electric-vehicle.

7. K. C. Colwell, "How Long Does It Take to Charge an Electric Vehicle?" *Car and Driver*, May 22, 2020, https://www.caranddriver.com/shopping-advice/a32600212/ev-charging-time/.

8. "Electric Car Chargers Will Determine America's Green Future," Bloomberg News, June 1, 2020, https://www.bloomberg.com/news/features/2020-06-01/electric-car-chargers-will-determine-america-s-green-future?cmpid=BBD060520_GREENDAILY&utm_medium=email&utm_source=newsletter&u=true&sref=x fyiavTX.

9. Justin Pritchard, "Is Your Dealership Equipped to Service Your Electric/Hybrid Car," *Chronicle Herald*, May 24, 2019, https://www.thechronicleherald.ca/wheels/is-your-dealership-equipped-to-service-your-electrichybrid-car-315420/.

10. Partly as a consequence of its commitment to an electric vehicle while most auto manufacturers made only tepid commitments, largely for compliance reasons, Tesla has stolen a march on an entire industry. In 2019, the company is estimated to have sold 81% of all EVs in the United States. See Zachary Shahan, "Tesla Gobbled Up 78% of US Electric Vehicle Sales in 2019," CleanTechnica, January 16, 2020, https://cleantechnica.com/2020/01/16/tesla-gobbled-up-81-of-us-electric-vehicle-sales-in-2019/.

11. As of the end of 2019, California's cumulative EV sales stood at 700,000, just under half of cumulative U.S. sales of 1,451,000 ("Sales Dashboard," Veloz, accessed November 24, 2020, https://www.veloz.org/sales-dashboard/). Many auto companies sell EVs simply to meet compliance goals—a function of a California requirement that require automakers to sell specific numbers of zero-emissions vehicles in the state if they want to be able to market there at all. See Larry Hall, "What Is a Compliance Car?" Treehugger, last updated October 30, 2020, https://www.thoughtco.com/what-is-a-compliance-car-85648.

12. In 2020, EV sales have declined considerably since COVID-19 sunk its invisible teeth into the American economy, crushed oil prices, wreaked havoc on global automotive supply chains, and shrunk the customer's wallet. That said, the impacts are expected to be only temporary, as manufacturers begin to introduce the models American drivers appear to really want that go beyond simple sedans. These include the new Ford Mustang, whose First Edition sold out, and then of course the all-important SUVs and pickup trucks. A slew of these models will be rolling off assembly lines in the coming years. Pickups include Tesla's uniquely shaped Cybertruck, a promising vehicle from startup Rivian (fortified with $1.3 billion of investment, including a huge chunk from Amazon), and the Ford 150.

13. Rev Up: A Nationwide Study of the Electric Vehicle Shopping Experience," Sierra Club, November 2019, https://www.sierraclub.org/sites/www.sierraclub.org/files/press-room/RevUpReportFinal.pdf. Not surprisingly, Tesla was cited as offering the best shopping experience, with a score of 4.5 out of 5.

14. Michael Hagerty, Sanem Sergici, and Long Lam, "Getting to 20 Million EVs by 2030: Opportunities for the Electricity Industry in Preparing for an EV Future," Brattle Group, June 2020, https://brattlefiles.blob.core.windows.net/files/19421_brattle_-_opportunities_for_the_electricity_industry_in_ev_transition_-_final.pdf.

15. Evannex, "Here's Seven Reasons Why Electric Vehicles Will Kill the Gas Car," InsideEVs, September 22, 2018, https://insideevs.com/news/340502/heres-seven-reasons-why-electric-vehicles-will-kill-the-gas-car/. Tesla claims its drivetrain has 17 moving parts, compared with roughly 200 in a standard vehicle. See John O'Dell, "10 Things That Make Electric-Car Maker Tesla Special," MarketWatch, August 23, 2016, https://www.marketwatch.com/story/10-things-that-make-the-tesla-a-great-car-2016-08-19.

16. The same day I was writing this chapter, I received an email from the Hyundai dealership from which we leased our Ioniq, informing me that the car was due

for its 3,750-mile service (in fact, the only recommended maintenance is tire rotation at 7,500 miles). A week ago I received a separate email from Hyundai North America with my "Monthly Vehicle Health Report," showing me exactly how many miles were on the odometer (3,131), since the car is connected to Hyundai via Bluelink.

17. Hagerty, Sergici, and Lam, "Getting to 20 Million EVs by 2030," Brattle Group.

18. "All-Electric Vehicles," U.S. Department of Energy, Office of Energy Efficiency and Renewable Energy, accessed November 24, 2020, https://www.fueleconomy.gov/feg/evtech.shtml.

19. This is just one of many online videos (warning—people are prone to swearing in response to the rapid acceleration) demonstrating passengers' reaction to "insane mode": "Tesla P85D Insane Mode Launch Reactions Compilation: Explicit Version with Brooks Weisblat," YouTube, January 25, 2015, https://www.youtube.com/watch?v=LpaLgF1uLB8. My son's Uber driver put his car into ludicrous mode on the highway (not recommended), and he was duly impressed.

20. The Taycan turbo can accelerate from 0 to 60 in 2.4 seconds. According to *Car and Driver*, that puts it on par with the Bugatti Veyron. It is bested only by Lamborghini's Huracan Performante and the Porsche 911 Spyder. Connor Hoffman, "Porsche Taycan Turbo S Is the Third-Quickest Car We've Ever Tested," *Car and Driver*, January 29, 2020, https://www.caranddriver.com/reviews/a30688949/2020-porsche-taycan-turbo-s-testing-acceleration-zero-to-60/.

21. Brian Silvestro, "A Used Chevy Spark EV Is the Discount Sleeper Hot Hatch You Didn't Know You Needed," Road & Track, December 24, 2018, https://www.roadandtrack.com/car-culture/buying-maintenance/a25632693/chevy-spark-ev-used-car-deal/.

22. Iqtidar Ali, "Battery Expert: Tesla Model 3 Has 'Most Advanced Large Scale Lithium Battery Ever Produced,'" Evannex, June 27, 2018, https://evannex.com/blogs/news/tesla-s-battery-pack-is-both-mysterious-and-alluring-work-in-progress.

23. In traditional cars, brakes are based on friction, with the kinetic energy of the vehicle converted into useless waste heat.

24. Micah Toll, "Regenerative Braking: How It Works and Is It Worth It in Small EVs?" Electrek, April 24, 2018, https://electrek.co/2018/04/24/regenerative-braking-how-it-works/.

25. Ibid.

26. "Battery Pack Prices Fall as Market Ramps Up with Market Average at $156/kWh in 2019," BloombergNEF, December 3, 2019, https://about.bnef.com/blog/battery-pack-prices-fall-as-market-ramps-up-with-market-average-at-156-kwh-in-2019/.

27. Antuan Goodwin, "2012 Nissan Leaf SL Review: Nissan Builds an Electric Car for the Rest of Us," Road Show by CNET, November 13, 2012, https://www.cnet.com/roadshow/reviews/2012-nissan-leaf-review/.

28. "2020 Nissan LEAF Range, Charging & Battery," Nissan.

29. Sebastian Blanco, "2020 Hyundai Ioniq EV Will Come with 170 Miles of Range, 133 MPGe," *Car and Driver*, November 8, 2019, https://www.caranddriver.com/news/a29740868/2020-hyundai-ioniq-range-mpge/. *Car and Driver* indicates my 2019 Ioniq gets 4.0 miles per kWh, while the 2020 Tesla Model 3 Long Range (75 kWh!—that's 2.5 days of average U.S. electricity consumption) gets 3.8 miles, the Model 3 Standard Range Plus gets 4.2, and the 2020 Chevy Bolt and Nissan Leaf get 3.1 and 3.4 miles per kWh respectively.

30. David Z. Morris, "Today's Cars Are Parked 95% of the Time," Yahoo Finance, March 13, 2016, https://finance.yahoo.com/news/today-cars-parked-95-time-210616765.html.

31. "Driver's Checklist: A Quick Guide to Fast Charging," ChargePoint, accessed November 24, 2020, https://www.chargepoint.com/files/Quick_Guide_to_Fast_Charging.pdf.

32. Ibid.

33. "Charging at Home," U.S. Department of Energy, Office of Energy Efficiency and Renewable Energy, accessed November 24, 2020, https://www.energy.gov/eere/electricvehicles/charging-home.

34. "Charge on the Road," Tesla, accessed November 24, 2020, https://www.tesla.com/supercharger.

35. "Introducing V3 Supercharging," Tesla, March 6, 2019, https://www.tesla.com/blog/introducing-v3-supercharging. Tesla recently introduced a technology that facilitates that new charging rate, so that cars can gain 75 miles of range in five minutes and charge at rates of up to 1,000 miles of range per hour. See Andrew Hawkins, "The Electric Vehicle Cannonball Run Record Was Broken Twice in One Month," The Verge, August 5, 2019, https://www.theverge.com/2019/8/5/20751975/ev-cannonball-run-record-broken-twice-2019. These fast rates of charge have led some enthusiasts to attempt the electric Cannonball Run—setting the record for the fastest electric-powered journey across the United States, from New York to California. A Swiss family set the record of 48 hours and 10 minutes in July 2019, and was promptly eclipsed the following month by an American duo driving a Tesla Model 3 that posted a time of 45 hours and 16 minutes.

36. "Electrify America to Open 400th Public Electric Vehicle Charging Station," Electrify America, February 4, 2020, https://media.electrifyamerica.com/en-us/releases/90.

37. Dave Venderwerp, "Porsche Taycan Can Charge Far More Quickly Than Tesla Model S," *Car and Driver*, February 13, 2020, https://www.caranddriver.com/reviews/a30894056/porsche-taycan-fast-charging-tesla-model-s/. It also notes that peak charging rates are only an indicator and do not translate to real-world experience, since they occur for only short durations, when the battery is nearly empty.

38. "Electric Vehicles at Scale, Phase I Analysis: High EV Adoption Impacts on the Western U.S. Power Grid," Pacific Northwest National Laboratory, July 2020, https://www.pnnl.gov/sites/default/files/media/file/EV-AT-SCALE_1_IMPACTS_final.pdf. This report on the potential impacts of EVs on the grid states, "We recognize that 2000 kW charging stations do not currently exist, but we assumed that by 2028 extreme fast charging may become available."

39. Fred Lambert, "NIO Might Have Figured Out Battery Swap for Electric Cars as It Completes 500,000 Swaps," Electrek, June 2, 2020, https://electrek.co/2020/06/02/nio-battery-swap-electric-cars-completes-500000-swaps/.

40. Jim Gorzelany, "What It Costs to Charge an Electric Vehicle," MyEV.com, accessed November 24, 2020, https://www.myev.com/research/ev-101/what-it-costs-to-charge-an-electric-vehicle. A May 2019 article in MyEV.com noted that largest network owner ChargePoint permits the owner of Level 2 chargers to set prices, while rival Blink levies between $0.04 and $0.06 per minute, or $0.39 to $0.79 per kWh. Level 3 is faster, but also more expensive. For example, the author of that piece paid $0.29 per minute for DC Fast Charging near Chicago at an EVgo station (which would have cost $0.25 per minute for EVgo subscribers.) A 25-minute session adding close to 50 miles of range to a VW eGolf cost $7.25, approximately $3.62 for 25 miles—more expensive than fueling a gas-powered VW Golf over the same miles.

41. "Charging Station Map of North America," Plugshare, accessed November 24, 2020, https://www.plugshare.com/EV-Charging-Networks-North-America.html. Plugshare, a company offering an app to locate charging stations, lists 15 companies in the United States and Canada. They vary tremendously in size, from ChargePoint, with tens of thousands of mostly Level 2 stations, to Electrify America, with a few hundred stations, but most of them DC high-speed Level 3 installations.

42. "Interoperability of Public Electric Vehicle Charging Infrastructure," Electric Power Research Institute (EPRI), August 2019, https://www.eei.org/issuesandpolicy/electrictransportation/Documents/Final%20Joint%20Interoperability%20Paper.pdf.

43. In addition, work needs to be done to create more flexibility between charging stations and networks; today, charging stations must link back to their networks, but they are often committed to the single company supporting its network. An open standards approach would allow owners and operators to switch between networks over time without having to buy new stations.

44. "Interoperability of Public Electric Vehicle Charging Infrastructure," EPRI. As EPRI observes, "The lack of a single accepted standard for DC charging for light duty EVs increases operational complexity and costs, and can lead to customer confusion as public DC fast charging expands."

45. The good news is that there is progress being made in that arena. For example, there is an open charge point protocol (OCPP), and a network standard from

Europe that is now gaining traction in the United States, with a growing number of U.S. networks accepting the OCPP.

46. "Waitlist," ChargePoint, accessed November 29, 2020, https://www.chargepoint.com/products/waitlist/. Companies like ChargePoint have ways for their subscribers to address this challenge, such as its Waitlist option. This allows drivers to use their cellphone to "get in line" if all charging points are occupied. Drivers already connected get messages when their EVs are charged, asking them to move, and once a station is free, the waitlisted driver is notified and the station is held open until they connect. ChargePoint even allows drivers to use automatic scheduling that puts them on a daily waitlist schedule—a feature helpful in busy workplace charging stations. However, this is of no benefit to others outside the company network.

47. James Taylor fans will immediately recognize that phrase.

48. "MassDOT: Framingham Electrical Vehicle Charging Station Now Activated," MassDOT Blog, November 17, 2017, https://blog.mass.gov/transportation/massdot-highway/massdot-framingham-electrical-vehicle-charging-station-now-activated/. The free charging policy came to an end on January 7, 2019; see Plug and Play EV, "Free EV Charging: End of Electric Vehicle Perks on the Mass Pike a Good Sign?" YouTube, January 9, 2019, https://www.youtube.com/watch?v=ieVTDAXA_EU.

49. "Charging Locations," EVgo, accessed November 24, 2020, https://www.evgo.com/charging-locations/. It now costs 30 cents a minute to get a 50 kW charge.

50. The Leaf's standard 40 kWh battery offers 149 miles of range. My interview subject's Leaf has a 62 kWh battery good for an EPA-rated 226 miles. Nissan states on its website ("2020 Nissan LEAF Range, Charging & Battery," Nissan USA, accessed December 6, 2020, https://www.nissanusa.com/vehicles/electric-cars/20-leaf/features/range-charging-battery.html) that one hour of charging at the 50 kW rate will get your battery charged to 80%.

51. "Electric Vehicle Sales to Fall 18% in 2020 but Long-Term Prospects Remain Undimmed," BloombergNEF, May 19, 2020, https://about.bnef.com/blog/electric-vehicle-sales-to-fall-18-in-2020-but-long-term-prospects-remain-undimmed/.

52. Ibid.

53. HEVO is one of a number of wireless charging companies entering this space. For example, competitor Lumen has recently achieved certification from UL, and BMW introduced wireless pad charging for one of its models in 2018. See James Hetherington, "BMW vs. Tesla: Elon Musk Beaten to Punch for Killer New Feature," *Newsweek*, May 29, 2018, https://www.newsweek.com/bmws-new-tesla-killing-electric-car-feature-wireless-charging-946899.

54. Mia Yamauchi, "How Wireless EV Charging Works for Tesla Model S," Plugless, accessed November 24, 2020, https://www.pluglesspower.com/learn/wireless-ev-charging-works-tesla-model-s/.

55. 240 volt is the same as the voltage for a residential washer and dryer.

56. Piceras, "*2001: A Space Odyssey* Docking Sequence (Space Ballet) set to Nachthelle," YouTube, January 21, 2012, https://www.youtube.com/watch?v=48rh85d4yTI.

57. "Guidehouse Insights Report Shows Market Volume for Wireless EV Charger Deployments Is Projected to Near $3 Billion by 2030," Businesswire, December 8, 2020, https://www.businesswire.com/news/home/20201208005201/en/Guidehouse-Insights-Report-Shows-Market-Volume-for-Wireless-EV-Charger-Deployments-Is-Projected-to-Near-3-Billion-by-2030.

58. Florian Knobloch, Steef V. Hanssen, Aileen Lam, Hector Pollitt, Pablo Salas, Unnada Chewpreecha, Mark A. J. Huijbregts, and Jean-Francois Mercure, "Net Emission Reductions from Electric Cars and Heat Pumps in 59 World Regions Over Time," *Nature Sustainability* 3 (March 2020), 437–447, https://www.nature.com/articles/s41893-020-0488-7.

59. Dale Hall and Nic Lutsey, "Effects of Battery Manufacturing on Electric Vehicle Life-Cycle Greenhouse Gas Emissions," International Council on Clean Transportation, February 2018, https://theicct.org/sites/default/files/publications/EV-life-cycle-GHG_ICCT-Briefing_09022018_vF.pdf.

60. "Lithium-Ion Battery Safety Issues for Electric and Plug-In Hybrid Vehicles," National Highway Traffic Safety Administration, October 2017, https://www.nhtsa.gov/sites/nhtsa.dot.gov/files/documents/12848-lithiumionsafetyhybrids_101217-v3-tag.pdf.

61. Jesse Roman, "Stranded Energy," *National Fire Protection Association Journal*, January 1, 2020, https://www.nfpa.org/News-and-Research/Publications-and-media/NFPA-Journal/2020/January-February-2020/Features/EV-Stranded-Energy.

62. Kevin Forestieri, "NTSB: Tesla's Autopilot Steered Model X into Highway Median, Causing Fatal Mountain View Crash," *Almanac News*, February 27, 2020, https://www.almanacnews.com/news/2020/02/27/ntsb-teslas-autopilot-steered-model-x-into-highway-median-causing-fatal-mountain-view-crash. After a two-year investigation, the National Transportation Safety Board determined the driver, an Apple engineer, was likely playing a video game during the time of the crash, as his autopilot steered the car straight into the median. Information from the car's system indicated the driver's hands were absent from the wheel for roughly one-third of his half-hour commute, with no attempt to alter the course of the vehicle as it approached the barrier at roughly 70 miles per hour.

63. According to one source ("Arc Flash: What Is It, Why Does It Happen, and How Can You Avoid It?" D&F Liquidators, accessed November 25, 2020, https://www.dfliq.net/blog/arc-flash-happen-can-avoid/), "arc flash" involves movement of undesirable electric discharge "through the air from one voltage phase to another, or to ground." This is a rapid temperature and pressure increase in the air "between electrical conductors, causing an explosion known as an arc blast." In a worst case, it can result in pressure waves and flying shrapnel.

64. In order to continue burning, a fire needs three elements: 1) a source of ignition, 2) fuel, and 3) oxygen. A lithium-ion cell provides all three, and therefore it cannot be extinguished. But it can be kept from propagating by reducing the temperature so that the next cell does not catch on fire in turn. That's why firefighters use huge amounts of water—simply to cool the cells. If you can keep propagation from occurring, the cell on fire will relatively quickly burn itself out.

65. Roman, "Stranded Energy," *National Fire Protection Association Journal.*

66. "NREL Analysis Explores Demand-Side Impacts of a Highly Electrified Future," National Renewable Energy Laboratory, July 9, 2018, https://www.nrel.gov/news/program/2018/analysis-demand-side-electrification-futures.html.

67. "To Start New Dendro Drive House Service from 2019," Mitsubishi Motors Corporation, March 5, 2019, https://www.mitsubishi-motors.com/en/innovation/motorshow/2019/gms2019/dendo/.

68. Blagojce Krivevski, "Nissan Accepting Electricity as Payment for Parking," Electric Cars Report, July 31, 2020, https://electriccarsreport.com/2020/07/nissan-accepting-electricity-as-payment-for-parking/.

69. Peter Kelly-Detwiler, "Driving Change: Transportation and Electric Utility Industries Will Soon Collide—In a Good Way," Forbes.com, October 15, 2019, https://www.forbes.com/sites/peterdetwiler/2019/10/15/driving-change-transportation-and-electric-utility-industries-will-soon-collide--in-a-good-way/#19b3ff9b47ca.

70. Steve Hanley, "Volkswagen Bets on Vehicle-to-Grid Technology, UL Approves First V2G Certification," CleanTechnica, March 13, 2020, https://cleantechnica.com/2020/03/13/volkswagen-bets-on-vehicle-to-grid-technology-ul-approves-first-v2g-certification/.

71. "Autobidder," Tesla, accessed November 25, 2020, https://www.tesla.com/support/autobidder.

72. "A Million-Mile Battery from China Could Power Your Electric Car," Bloomberg News, June 7, 2020, https://www.bloomberg.com/news/articles/2020-06-07/a-million-mile-battery-from-china-could-power-your-electric-car?sref=xfyiavTX.

73. Roberto Baldwin, "China's CATL Has a Million-Mile EV Battery Pack Ready to Go," *Car and Driver*, June 8, 2020, https://www.caranddriver.com/news/a32801823/million-mile-ev-battery-pack-revealed/.

74. "Volkswagen Plans to Tap Electric Car Batteries to Compete with Power Firms," Reuters, March 12, 2020, https://www.reuters.com/article/us-volkswagen-electric-energy/volkswagen-plans-to-tap-electric-car-batteries-to-compete-with-power-firms-idUSKBN20Z2D5.

75. Sven Gustafson, "Your EV's Electricity Will Cover Parking Fee at Nissan's New Pavilion," Autoblog, August 8, 2020, https://www.autoblog.com/2020/08/08/ev-electricity-pay-parking-nissan-pavilion/.

76. Hagerty, Sergici, and Lam, "Getting to 20 Million EVs by 2030," Brattle Group.

77. I. Wagner, "Number of Vehicles in Operation in the United States between 1st Quarter 2016 and 4th Quarter 2019," Statista, May 14, 2020, https://www.statista.com/statistics/859950/vehicles-in-operation-by-quarter-united-states/.

78. Hagerty, Sergici, and Lam, "Getting to 20 Million EVs by 2030," Brattle Group.

79. Ibid.

CHAPTER 14

1. Kim Zetter, "Inside the Cunning, Unprecedented Hack of Ukraine's Power Grid," Wired, March 3, 2016, https://www.wired.com/2016/03/inside-cunning-unprecedented-hack-ukraines-power-grid/.

2. Joe Slowik, "Crashoverride: Reassessing the 2016 Ukraine Electric Power Event as a Protection-Focused Attack," Dragos, August 15, 2019, https://www.dragos.com/wp-content/uploads/CRASHOVERRIDE.pdf?__hstc=8780330.ddeeedb800a5846837211dbaa372db99.1588193020210.1588348690035.1588874747409.3&__hssc=8780330.1.1588874747409. In fact, closer analysis revealed that a far more widespread outage was the intended outcome in the largely failed effort—affecting hundreds of devices, rather than the one transformer impacted and potentially "orders of magnitude larger" than the 2015 incident. While the initial attempt failed, follow-up activities included deployment of a wiper that would slow a return to normal on restoration of SCADA systems and reduce overall situational awareness. Most importantly, this activity was then trailed by an attempted denial-of-service attack on four Siemens protective relays specifically designed to protect equipment from damage. It seems that the ultimate goal was to create an unstable environment during the restoration process that may have led to a more widespread outage with significant damage to equipment for as long as a month or more.

3. "North American Electric Cyber Threat Perspective," Dragos, January 2020, https://www.dragos.com/wp-content/uploads/NA-EL-Threat-Perspective-2019.pdf.

4. Ibid. While Dragos takes pains not to attribute specific groups to nation-states, the four most active players are China, Iran, North Korea, and Russia.

5. "United States v. Ahmad Fathi, Hamid Firoozi, Amin Shokohi, Sadegh Ahmadzadegan Omid Ghaffarinia, Sina Keissar, and Nader Saedi: Indictment," U.S. District Court, Southern District of New York, accessed November 27, 2020, https://www.justice.gov/usao-sdny/file/835061/download. (There is another—substantially larger—240-foot Bowman Dam in Oregon they may have been targeting; see "Arthur R. Bowman Dam," U.S. Bureau of Reclamation, accessed November 27, 2020, https://www.usbr.gov/projects/index.php?id=45.)

6. Rebecca Smith, "Russian Hackers Reach U.S. Utility Control Rooms, Homeland Security Officials Say," Wall Street Journal, July 23, 2018, https://www.

wsj.com/articles/russian-hackers-reach-u-s-utility-control-rooms-homeland-security-officials-say-1532388110.

7. Michael Raggi, "LookBack Forges Ahead: Continued Targeting of the United States' Utilities Sector Reveals Additional Adversary TTPs," Proofpoint, September 23, 2019, https://www.proofpoint.com/us/threat-insight/post/lookback-forges-ahead-continued-targeting-united-states-utilities-sector-reveals.

8. "Lesson Learned: Risks Posed by Firewall Firmware Vulnerabilities," North American Electric Reliability Corporation, September 4, 2019, https://www.eenews.net/assets/2019/09/06/document_ew_02.pdf.

9. "North American Electric Cyber Threat Perspective," Dragos.

10. "Triton 2.0 & The Future of OT Cyber-Attacks," DarkTrace Industrial, accessed November 27, 2020, https://img.en25.com/Web/PentoniNET/%7Bec9a8539-a4f8-4e72-bede-0d77e33cceda%7D_A2020124-DarkTrace-Triton_2.0__the_Future_of_OT_Cyber-Attacks.pdf.

11. Donald J. Trump, "Executive Order on Securing the United States Bulk-Power System," the White House, May 1, 2020, https://www.whitehouse.gov/presidential-actions/executive-order-securing-united-states-bulk-power-system/. There is still some uncertainty as to what that actually means, and how it would be applied. In July, the U.S. Department of Energy issued a request for information, seeking comments, and announced it expected to issue a notice of proposed rulemaking later in 2020. See "Securing the United States Bulk-Power System Executive Order," U.S. Department of Energy, Office of Electricity, last updated November 1, 2020, https://www.energy.gov/oe/bulkpowersystemexecutiveorder.

12. Alexander D. Schlichting, "Assessment of Operational Energy System Cybersecurity Vulnerabilities," MITRE Corporation, January 2018, https://www.mitre.org/sites/default/files/publications/pr_18-1118-assessment-operational-energy-system-cybersecurity-vulnerabilities.pdf.

13. J. D. Taft and R. Huang, "Distribution Storage Networks," GRID Modernization Laboratory Consortium, June 2017, https://gmlc.doe.gov/sites/default/files/resources/Distribution%20Storage%20Networks%20v0.3_GMLC.pdf.

14. "Cumulative Installed PV—As of Mar. 31, 2020," Hawaiian Electric, accessed November 27, 2020, https://www.hawaiianelectric.com/documents/clean_energy_hawaii/clean_energy_facts/pv_summary_1Q_2020.pdf.

15. John Weaver, "Residential Solar Plus Storage Is Taking Over Hawaii's Grid," PV Magazine, September 3, 2019, https://pv-magazine-usa.com/2019/09/04/hawaiian-homes-becoming-even-more-important-to-the-power-grid/.

16. "Shifted Energy to Equip 2,400 Water Heaters in Hawaii with Grid-Interactive Technology to Create Virtual Power Plant," Shifted Energy, October 1, 2019, https://www.globenewswire.com/news-release/2019/10/01/1923461/0/en/Shifted-Energy-to-Equip-2-400-Water-Heaters-in-Hawaii-with-Grid-Interactive-Technology-to-Create-Virtual-Power-Plant.html.

17. "Medium and Heavy Duty Electric Vehicle and Charging Infrastructure Cyber Security Baseline Reference Document," National Motor Freight Traffic

Association, May 30, 2018, http://www.nmfta.org/documents/hvcs/MDH-DEV%20CI%20Cyber%20Security%20v1%202%201%20complete.pdf?v=1.

18. Christian Vasquez, "'Major Vulnerability': EV Hacks Could Threaten Power Grid," E&E News, June 17, 2020, https://www.eenews.net/stories/1063401375.

19. Kelsey Misbrener, "Smart Inverters Redefine Relationship between DERs and the Grid," Solar Power World, March 12, 2019, https://www.solarpower-worldonline.com/2019/03/smart-inverters-redefine-relationship-ders-grid/.

20. Kelsey Misbrener, "Cyberattacks Threaten Smart Inverters, But Scientists Have Solutions," Solar Power World, April 30, 2019, https://www.solarpow-erworldonline.com/2019/04/cyberattacks-threaten-smart-inverters-but-scientists-have-solutions/.

21. "FERC Order No. 2222: A New Day for Distributed Energy Resources," Federal Energy Regulatory Commission (FERC), September 17, 2020, https://www.ferc.gov/sites/default/files/2020-09/E-1-facts.pdf.

22. "'Something Astounding Just Happened': Enphase's Grid-Stabilizing Collaboration with Hawaiian Electric," Enphase, March 11, 2015, https://enphase.com/en-us/stories/%E2%80%98something-astounding-just-happened%E2%80%99-enphase%E2%80%99s-grid-stabilizing-collaboration-hawaiian.

23. "CME Week: The Difference Between Flares and CMEs," National Aeronautics and Space Administration (NASA), September 21, 2014, https://www.nasa.gov/content/goddard/the-difference-between-flares-and-cmes.

24. Ibid. NASA indicates that through a telescope, a flare simply appears as a very bright light, while a coronal mass ejection is visible as "enormous fans of gas swelling into space."

25. "In March 1989, Québec Experienced a Blackout Caused by a Solar Storm," Hydro Québec, accessed November 27, 2020, http://www.hydroquebec.com/learning/notions-de-base/tempete-mars-1989.html.

26. "March 13, 1989 Geomagnetic Disturbance Report," North American Electric Reliability Council, July 9, 1990, https://www.nerc.com/pa/Stand/Geomagnetic%20Disturbance%20Resources%20DL/NERC_1989-Quebec-Disturbance_Report.pdf.

27. "Reliability Standards for Geomagnetic Disturbances," FERC, May 16, 2013, https://www.ferc.gov/industries/electric/indus-act/reliability/cybersecurity/ferc_meta-r-319.pdf.

28. Christopher Klein, "A Perfect Solar Superstorm: The 1859 Carrington Event," History.com, last updated August 22, 2018, https://www.history.com/news/a-perfect-solar-superstorm-the-1859-carrington-event.

29. John Kemp, "Time to Be Afraid: Preparing for the Next Big Solar Storm," Reuters, July 24, 2014, https://www.reuters.com/article/us-electricity-solar-storms-kemp/time-to-be-afraid-preparing-for-the-next-big-solar-storm-kemp-idUSKBN0FU20Q20140725. The Dst index measures how severely the Earth's magnetic field is affected by space weather.

30. "Near Miss: The Solar Superstorm of July 2012," NASA, July 23, 2014, https://science.nasa.gov/science-news/science-at-nasa/2014/23jul_superstorm.

31. According to the 2019 Electric Power Institute report on the subject ("High-Altitude Electromagnetic Pulse and the Bulk Power System: Potential Impacts and Mitigation Strategies," April 29, 2019, https://www.epri.com/research/products/3002014979), the International Electrotechnical Commission (IEC) defines three HEMP hazard fields based on their distinct characteristics and time scales: 1) The early time component (E1 EMP) consisting of an intense, short-duration electromagnetic pulse characterized by a rise time of 2.5 nanoseconds and amplitude on the order of tens of kV/m (up to 50 kV/m at the most severe location on the ground); 2) The intermediate time component (E2 EMP), which is considered an extension of E1 EMP and has an electric field pulse amplitude on the order of 0.1 kV/m and duration of 1 microsecond to approximately 10 milliseconds; 3) The late time component (E3 EMP), which is a very low frequency (below 1 Hz) pulse with amplitude on the order of tens of V/km with duration of one second to hundreds of seconds. E3 EMP is often compared with severe geomagnetic disturbance (GMD) events; however, the intensity of E3 EMP can be orders of magnitude more severe, and E3 EMP is much shorter in duration than GMD events, which can last for several days.

32. For example, the North American Aerospace Command (NORAD) and U.S. Northern Command announced plans to move resources including communications equipment to the EMP-hardened Cheyenne Mountain complex. See "Strategic Primer: Electromagnetic Threats," American Foreign Policy Council, Winter 2018, https://www.afpc.org/uploads/documents/EMP%20Primer%20-final.pdf.

33. "Report of the Commission to Assess the Threat to the United States from Electromagnetic Pulse (EMP) Attack," Critical National Infrastructures, April 2008, http://www.empcommission.org/docs/A2473-EMP_Commission-7MB.pdf.

34. "EMP Task Force: Strategic Recommendations," North American Electric Reliability Corporation, November 5, 2019, https://www.nerc.com/pa/Stand/EMP%20Task%20Force%20Posting%20DL/NERC_EMP_Task_Force_Report.pdf. The report observes that "the disruptive influence of an EMP event seems likely to span across the full spectrum of power system assets, including the transmission system, the distribution system, the protections and controls hardware, and the command and control infrastructure relied upon to monitor and maintain the power system in a stable operating state."

35. "Strategic Primer: Electromagnetic Threats," American Foreign Policy Council. In late 2017, North Korea discussed a possible atmospheric nuclear test.

36. Donald J. Trump, "Executive Order on Coordinating National Resilience to Electromagnetic Pulses," the White House, March 26, 2019, https://www.whitehouse.gov/presidential-actions/executive-order-coordinating-national-resilience-electromagnetic-pulses/.

37. "2013 Attack on Metcalf, California Power Grid Substation Committed by 'an Insider': DHS," Homeland Security Newswire, October 19, 2015, http://www.homelandsecuritynewswire.com/dr20151019-2013-attack-on-metcalf-california-power-grid-substation-committed-by-an-insider-dhs. In that instance, the attackers were never identified and there was some speculation that it may have been an inside job, the result of disgruntled employees.

38. A 2012 National Academy of Sciences report on the topic of grid security observed that "Substations and the large high-voltage transformers they contain are especially vulnerable, as are some transmission lines where the destruction of a small number of towers could bring down many kilometers of line. Terrorist attacks on multiple-line transmission corridors could cause cascading blackouts." See National Research Council, *Terrorism and the Electric Power Delivery System* (Washington, DC: National Academies Press, 2012), chapter 2 excerpted at https://www.nap.edu/read/12050/chapter/2#2.

39. "Physical Security of the U.S. Power Grid: High-Voltage Transformer Substations," EveryCRSReport.com, July 2, 2015, https://www.everycrsreport.com/reports/R43604.html. A 2014 *Wall Street Journal* article cited an unreleased report from FERC indicating that if just nine critical substations across the country were taken out, "The U.S. could suffer a coast-to-coast blackout." Given the difficulty of procuring and moving these transformers, recovery could take weeks, if not months. See Rebecca Smith, "U.S. Risks National Blackout from Small-Scale Attack: Federal Analysis Says Sabotage of Nine Key Substations Is Sufficient for Broad Outage," *Wall Street Journal*, March 12, 2014, https://www.wsj.com/articles/SB10001424052702304020104579433670284061220.

40. "Disaster Preparedness: The Quest for Transformer Resilience," Power, April 1, 2018, https://www.powermag.com/disaster-preparedness-the-quest-for-transformer-resilience/.

41. Tim Williams, "Keeping NY's Power on During COVID19," Capitol Pressroom (podcast), WCNY, Apr. 15, 2020, http://www.wcny.org/apr-15-2020-keeping-nys-power-on-during-covid19/. In this radio interview, CEO Richard Dewey commented that as the news began filtering in during January, his organization began reviewing the plan, and eventually sequestered volunteers into two separate control rooms beginning March 23, renting trailers for accommodations. All employees were tested prior to lock-in, remained socially separate for first 14 days, and have access to mental health professionals. Dewey characterized the plan as "pretty useful" and commented, "we were pretty happy with the thoroughness of the plan" put in place a decade ago as part of the response to the bird flu epidemic.

42. "Grid Resilience: Priorities for the Next Administration," National Commission on Grid Resilience, 2020, https://gridresilience.org/wp-content/uploads/2020/09/NCGR-Report-2020-Full-v2.pdf. This August 2020 report from the National Commission on Grid Resilience lays out a number of recommendations for upgrading grid security (I was a coauthor of this report).

43. Ibid.

44. "Surviving a Catastrophic Power Outage: How to Strengthen the Capabilities of the Nation," The President's National Infrastructure Advisory Council, December 2018, https://www.cisa.gov/sites/default/files/publications/NIAC%20Catastrophic%20Power%20Outage%20Study_FINAL.pdf. The report notes, "Given the importance of this issue and the number of ongoing efforts, we request the National Security Council (NSC)—working with the lead agencies identified—provide a status update to the NIAC within nine months of the report's approval on how our recommendations are being implemented, progress being made on the ongoing initiatives, or any significant barriers to implementation." It is unclear whether this recommendation has been acted upon.

CHAPTER 15

1. "Progress Toward 100% Clean Energy in Cities and States Across the U.S." UCLA Luskin Center for Innovation, November 2019, https://innovation.luskin.ucla.edu/wp-content/uploads/2019/11/100-Clean-Energy-Progress-Report-UCLA-2.pdf.

2. Iulia Gheorghiu, "Green Mountain Power Turnaround Driven by Customer Obsession, Cultural Shift: CEO," Utility Dive, October 15, 2019, https://www.utilitydive.com/news/green-mountain-power-turnaround-driven-by-customer-obsession-cultural-shif/565031/.

3. "We're Proud to Be a Certified B Corporation," Green Mountain Power, December 1, 2014, https://greenmountainpower.com/proud-certified-b-corporation/. B Corporations are legally required to consider the impact of their decisions on their workers, customers suppliers, community, and the environment; see "A Global Community of Leaders," B Lab, accessed November 29, 2020, https://bcorporation.net.

4. In the case of Quebec's provincially owned utility, Hydro Québec, the development of its vast hydropower resources on aboriginal lands has not been without controversy.

5. Michael Liebreich, "Peak Emissions Are Closer Than You Think—and Here's Why," BloombergNEF, December 17, 2019, https://about.bnef.com/blog/peak-emissions-are-closer-than-you-think-and-heres-why/. Liebreich's piece is a highly readable and interesting description of how that thinking has changed, and he is one of the eminent thinkers in this arena.

6. Nelson Mojarro, "COVID-19 Is a Game-Changer for Renewable Energy. Here's Why," World Economic Forum, June 16, 2020, https://www.weforum.org/agenda/2020/06/covid-19-is-a-game-changer-for-renewable-energy/.

7. In recent years, existing nuclear plants have become increasingly uncompetitive to the point that some states, such as New York and Illinois, now specifically subsidize them with zero emission credits just to keep them running. See "Zero-

Emission Credits," Nuclear Energy Institute, April 2018, https://www.nei.org/resources/reports-briefs/zero-emission-credits.

8. A plant that sells 100 MW into the market may actually be built at 110 or 120 MW and deliver more energy when called upon, having the capability to flex up and down as the grid operator calls for it during a portion of its operating hours. Proving more "headroom" and "footroom" will allow a much larger percentage of solar to be integrated into the system. For a thorough description of how this might work, see "Investigating the Economic Value of Flexible Solar Power Plant Operation," Energy and Environmental Economics, October 2018, https://www.ethree.com/wp-content/uploads/2018/10/Investigating-the-Economic-Value-of-Flexible-Solar-Power-Plant-Operation.pdf.

9. "FERC Order No. 2222: A New Day for Distributed Energy Resources," FERC, September 17, 2020, https://www.ferc.gov/sites/default/files/2020-09/E-1-facts.pdf.

10. "Renewables," Kaua'i Island Utility Cooperative, accessed November 29, 2020, https://website.kiuc.coop/renewables.

11. Ibid. The region is cursed/blessed with the fact that it used to generate electricity from expensive diesel generators, so it had an expensive target at which to shoot. The utility noted that in 2019, it saved $3.8 million in diesel costs with its solar/storage resources.

12. "You're Flexible. So Are We! Flexible Path," CPS Energy, accessed November 29 2020, https://cpsenergy.com/flexiblepath.

13. Moore's Law refers to the concept that the number of transistors on a microchip doubles every two years, while the cost falls by 50%. Thus we can expect to have faster and more powerful machines and continually pay less.

14. "CPS Energy Seeks RFI to Identify Solar, Battery, and New Technology Solutions for FlexPOWER Bundle RFP," Peak Load Management Alliance, July 27, 2020, https://www.peakload.org/cps-rfp-7-28-2020; "FlexPower Bundle RFI," CPS Energy, accessed November 29, 2020, www.cpsenergy.com/rfi.

15. "Remarks of Chairman Neil Chatterjee on Order 2222," FERC, September 17, 2020, https://www.ferc.gov/news-events/news/remarks-chairman-neil-chatterjee-order-2222.

16. "Electricity Explained: Electricity Generation, Capacity, and Sales in the United States," EIA, accessed November 29, 2020, https://www.eia.gov/energy-explained/electricity/electricity-in-the-us-generation-capacity-and-sales.php.

17. "Lockheed Martin Pursuing Compact Nuclear Fusion Reactor Concept," Lockheed Martin, October 15, 2014, https://news.lockheedmartin.com/2014-10-15-Lockheed-Martin-Pursuing-Compact-Nuclear-Fusion-Reactor-Concept.

18. Joseph Trevithick, "The War Zone: Skunk Works' Exotic Fusion Reactor Program Moves Forward with Larger, More Powerful Design," The Drive, July 19, 2019, https://www.thedrive.com/the-war-zone/29074/skunk-works-exotic-fusion-reactor-program-moves-forward-with-larger-more-powerful-design.

19. David L. Chandler, "Validating the Physics behind the New MIT-Designed Fusion Experiment," MIT News, September 29, 2020, https://news.mit.edu/2020/physics-fusion-studies-0929.

20. Levin Nock, "Will Deep Decarbonization Include Nuclear? Possibly, with Many Caveats," Utility Dive, January 14, 2020, https://www.utilitydive.com/news/cant-never-have-too-much-fun-or-too-much-wind-and-solar/570295/.

21. Eugene Kiely, "Florida Surrogates Go Nuclear," FactCheck.org, August 18, 2014, https://www.factcheck.org/2014/08/florida-surrogates-go-nuclear/.

22. "Former SCANA Execs Accused by Feds in Ill-Fated V. C. Summer Nuclear Project," Power Engineering, February 28, 2020, https://www.power-eng.com/2020/02/28/former-scana-execs-accused-by-feds-in-ill-fated-v-c-summer-nuclear-project/#gref.

23. Alex Crees, "The Failed V. C. Summer Nuclear Project: A Timeline," Choose Energy, December 4, 2018, https://www.chooseenergy.com/news/article/failed-v-c-summer-nuclear-project-timeline/.

24. Sonal Patel, "How the Vogtle Nuclear Expansion's Costs Escalated," Power, September 24, 2018, https://www.powermag.com/how-the-vogtle-nuclear-expansions-costs-escalated/.

25. Matt Kempner, "Georgia Vogtle Nuclear Report: More Delays, $1B in Extra Costs, Flaws," Atlanta Journal-Constitution, June 8, 2020, https://www.ajc.com/news/local/georgia-vogtle-nuclear-report-more-delays-extra-costs-flaws/mBxlgXiDcf0SIaTFr0cZXL/.

26. A major investor is Fluor, an engineering and construction giant.

27. "The Carbon Free Power Project," NuScale Power, accessed November 29, 2020, https://www.nuscalepower.com/projects/carbon-free-power-project.

28. "NuScale's Affordable SMR Technology for All," NuScale Power, accessed November 29, 2020, https://www.nuscalepower.com/newsletter/nucleus-spring-2020/featured-topic-cost-competitive.

29. "August 2020 Investor Presentation," NextEra Energy, August 2020, http://www.investor.nexteraenergy.com/~/media/Files/N/NEE-IR/news-and-events/events-and-presentations/2020/8-10-2020/August%202020%20Investor%20Presentation%20vF.pdf.

30. In July 2019 I was asked to deliver a plenary session at the New Nuclear Capital conference. In that session, I presented information related to the pace of change in the renewables and energy storage industries, and the rapid decline in costs that would challenge the modular nuclear industry. In the following panel session, one of the executives on the dais pointed to me and said something to the effect of "He just told us we don't have a future."

31. "Six Things You Might Not Know About Hydrogen," Argonne National Laboratory, October 7, 2016, https://www.anl.gov/article/six-things-you-might-not-know-about-hydrogen.

32. Peter Maloney, "LADWP Embarks on Hydrogen Generation Project," American Public Power Association, December 18, 2019, https://www.publicpower.org/periodical/article/ladwp-embarks-hydrogen-generation-project.

33. Cornelia Lichner, "Electrolyzer Overview: Lowering the Cost of Hydrogen and Distributing Its Production," PV Magazine, March 26, 2020, https://pv-magazine-usa.com/2020/03/26/electrolyzer-overview-lowering-the-cost-of-hydrogen-and-distributing-its-productionhydrogen-industry-overview-lowering-the-cost-and-distributing-production/.

34. "Shell and GasUnie Plan to Build Massive Dutch Green Hydrogen Plant," Reuters, February 27, 2020, https://www.reuters.com/article/us-shell-gasunie-hydrogen/shell-and-gasunie-plan-to-build-massive-dutch-green-hydrogen-plant-idUSKCN20L1AV.

35. "Equinor Joins Europe's Biggest Green Hydrogen Project, the NortH2-Project," Equinor, December 7, 2020, https://www.equinor.com/en/news/20201207-hydrogen-project-north2.html.

36. John Parnell, "Ørsted Backs First Major Green Hydrogen Project Focused on Transport Sector," GTM, May 26, 2020, https://www.greentechmedia.com/articles/read/orsted-to-power-decarbonization-hub-for-land-sea-and-air-transport.

37. In early 2020, a Siemens vice president predicted green hydrogen could become the "new oil" (Stephen Lacey, "Could Green Hydrogen Become the 'New Oil'?" GTM, January 23, 2020, https://www.greentechmedia.com/articles/read/could-green-hydrogen-become-the-new-oil.

38. Karl-Erik Stromsta, "NextEra Energy to Build Its First Green Hydrogen Plant in Florida," GTM, July 24, 2020, https://www.greentechmedia.com/articles/read/nextera-energy-to-build-its-first-green-hydrogen-plant-in-florida.

ACKNOWLEDGMENTS

It would be hard to thank everybody who has contributed to this effort, but I will nonetheless make the effort.

Thanks to Joan Michelson, who for years simply would not give up on the concept of me publishing this book, and introduced me to my agent, Wendy Keller. To Wendy, who had faith in the concept and guided me to Prometheus Books. To Jake Bonar, my editor at Prometheus, who enthusiastically embraced the project.

Thanks to the LinkedIn community, and especially the dozens of you who volunteered to edit chapters of the book. Although I am sure there are still some errors there, you helped keep me from stepping in some major puddles. Of special note: Scott Bargerstock, Phil Chadwick, Joe Doetzl, Tom Feiler, Terry Harville, Mark Kleinginna, Mike Marchand, Imran Nourani, Joseph Pokalsky, James Preedy, Matt Rhodes, Andy Schlosser, Keith Scoles, Howard Smith, Kelly Speakes-Backman, Jim Steffes, Chris Sullivan, and (super editor) Peter Weigand. If I have failed to mention your contributions, please accept my apologies.

And finally, thank you to all of the individuals who patiently explained to me your piece of the enormously wide world of electricity. I am grateful to you for entrusting me with telling your stories. I sincerely hope I did them justice.

INDEX